Metallographic Etching

Metallographic and Ceramographic Methods for Revealing Microstructure

Günter Petzow

Max Planck Institute for Metals Research,
Institute for Materials Science,
Stuttgart, West Germany

Translators from the German version:
Rosemarie Koch and James A. Nelson

AMERICAN SOCIETY FOR METALS
Metals Park, Ohio 44073

Originally published in 1976 as
Metallographische Ätzen

Library of Congress Cataloging in Publication Data

Petzow, G.
Metallographic etching.
"An improved version of the 5th edition of . . . Metallographische Ätzen."
Contains bibliographic references.
Includes index.
1. Metallographic specimens, 2. Metals—Etching.
I. Title.
TN690.7P4713 1978 669′.95′028 78-8023
ISBN 0-87170-002-6

PRINTED IN THE UNITED STATES OF AMERICA

To Robert J. Gray and E. Daniel Albrecht, two distinguished contributors to the development of modern metallography, for their encouragement in preparing the English version of this book.

Preface

In the last few years, metallographic preparation techniques have been improved considerably. During this progress, the transition from manual and empirical methods to reproducible and automatic techniques—although not yet complete—has become a reality. Taking this into account, the modern preparation techniques and their functional interrelationships are first treated in the present book, then well-proven metallographic recipes for individual materials are listed. In addition to the classical materials, those special metals and alloys are treated which find their application in aerospace and nuclear engineering. Similarly, techniques and recipes for metal-ceramics and special ceramics are presented, because these materials also are investigated more and more by "metallographic"—or, better, "ceramographic"—methods.

Instructions for specimen preparation not only are numerous but also frequently contradict each other. Therefore, mainly the author's experience—rather than data in the literature—was used in compiling the preparation recipes in this book. Special attention was given to those procedures that appear particularly suitable for small or medium-size metallographic laboratories, which normally are not equipped with expensive and sophisticated instruments. The selected recipes are simple and well-proven in practice; complicated, seldom used procedures of poor reproducibility have been omitted.

The present English-language book, "Metallographic Etching," is an improved version of the 5th edition of the booklet "Metallographisches Ätzen" (in German), published in 1976. The proven grouping of the various recipes into certain classes of materials has been maintained, because it enables the ready exchange, variation, and combination of various procedures between and within certain groups of materials.

I wish to thank all those who assisted me in the preparation of this edition in English, particularly Rosemarie Koch and James A. Nelson for their translation services. For their critical reading of the manuscript I am indebted to Winfried J. Huppmann, Ian Yapp, and Alan Prince. And, finally, I would like to acknowledge the tedious task of typing the manuscript by Antonie Rohrbach, Gertrud Thede, and Inge Hörmann.

Stuttgart, 8 July 1977 GÜNTER PETZOW

Preface to the German Edition

Vorwort

Seit der letzten Auflage des Ätzheftes im Jahre 1957 hat sich die Situation auf dem Werkstoffgebiet weitgehend verändert. Das Aufkommen neuer und die Optimierung bekannter Technologien haben viele Werkstoffe hervorgebracht, deren Aufbau und Eigenschaften nicht zuletzt in den metallographischen Laboratorien in Forschung und Praxis aufgeklärt und überprüft werden müssen. Dem wurde auch die metallographische Präparationstechnik angepasst: Der Übergang von der handwerklichen Empirie zur reproduzierbaren, automatisierten Methode wurde, wenn auch nicht vollständig, so doch in erheblichem Ausmass vollzogen. Das alles musste sich natürlich auf Inhalt und Umfang dieser Auflage auswirken. So ist letzten Endes ein neues Buch unter neuem Titel entstanden, das nur noch teilweise mit der vorhergehenden Auflage zu vergleichen ist.

Neben einer alle modernen Methoden einschliessenden Beschreibung der Präparationstechniken wurde eine ganze Reihe von Werkstoffgruppen neu aufgenommen; angefangen von metallischen Hochleistungswerkstoffen im Flugzeug-, Reaktor- und Raketenbau bis hin zu den metallkeramischen und keramischen Sonderwerkstoffen, die in zunehmendem Mass lichtmikroskopisch untersucht werden. Die in den früheren Auflagen bewährte Anordnung der Rezepte in bestimmte Materialklassen wurde beibehalten, da sie die Übersicht über Austausch, Variation und Kombination innerhalb ähnlicher Werkstoffe erleichtert. Für das schnelle und gezielte Auffinden eines Rezeptes für eine bestimmte Anwendung wurde zusätzlich ein nach Werkstoffen geordnetes Stichwortverzeichnis angefügt.

Da die in der Literatur vorhandenen Angaben über die Schliffherstellung sehr zahlreich und teilweise widersprüchlich sind, wurde bei der Auswahl der Präparationsrezepte in erster Linie auf eigene Erfahrungen zurückgegriffen. Besonders beachtet wurde dabei ihre Eignung für kleine bis mittlere metallographische Laboratorien, in denen aufwendige Einrichtungen selten sind. Die ausgewählten Rezepte sind daher einfach und bewährt; komplizierte, sehr anfällige und wenig erprobte Angaben wurden nicht berücksichtigt.

Allen Damen und Herren, die mir bei der Fertigstellung des Ätzheftes geholfen haben, möchte ich herzlich danken: Insbesondere den Damen Karin Kuhn, Karin Exner und Gonde Kiessler, die mir bei der anfänglichen Durchsicht der Literatur und beim Lesen der Korrekturfahnen sehr behilflich waren. Für die kritischen Kommentare zu vielen Rezepten bin ich den Mitgliedern des Ausschusses für Metallographie der Deutschen Gesellschaft für Metallkunde allen voran Frau Grete Petrich sowie Herrn Dr. H. P. Hougardy und Herrn F. Spies vom Max-Planck-Institut für Eisenforschung sehr verbunden; ebenso Herrn Dr. Lipp vom Elektroschmelzwerk Kempten für die Korngrössenangaben der Schleif- und Poliermittel. Frau Dr. Angelica Schrader, die das Ätzheft im Jahre 1934 begründete und für alle vorhergehenden Ausgaben verantwortlich zeichnet, hat auch an der Fertigstellung dieser Auflage aktiven Anteil genommen.

Stuttgart, im November 1975 GÜNTER PETZOW

Contents

1. Technical Tips for Preparation of Metallographic Specimens—1
Specimen Sectioning—2
Mounting—3
Identification (Marking)—7
Grinding and Polishing—8
Mechanical Grinding and Polishing—9
Microtome Cutting—16
Electrolytic Grinding and Polishing—16
Chemical Polishing—20
Combination Polishing Methods—21
Automatic Grinding and Polishing—22
Evaluation of Polishing Methods—22
Cleaning—23
Etching—24
Optical Etching—25
Electrochemical (Chemical) Etching—25
Physical Etching—29
Specimen Storage—29
Reproducibility in Etching—30
Etching Nomenclature—31
Explanation of Etching Terms—31
Nondestructive Metallographic Testing—34

2. Preparation of Metals and Alloys—37
Ag Silver—37
Al Aluminum—39
Au Gold—43
Ir Iridium—43
Os Osmium—43
Pd Palladium—43
Pt Platinum—43
Rh Rhodium—43
Ru Ruthenium—43
Be Beryllium—46

Bi Bismuth—48
Sb Antimony—48
Cd Cadmium—50
In Indium—50
Tl Thallium—50
Co Cobalt—51
Cr Chromium—54
Mo Molybdenum—54
Nb Niobium—54
Re Rhenium—54
Ta Tantalum—54
V Vanadium—54
W Tungsten—54
Cu Copper—58
Fe Iron, Steel, Cast Iron—61
Ge Germanium—68
Se Selenium—68
Si Silicon—68
Te Tellurium—68
$A_{III}B_V$ and $A_{II}B_{VI}$ Compounds—68
Hf Hafnium—70
Zr Zirconium—70
Hg Mercury Alloys (Amalgamates)—71
Mg Magnesium—72
Mn Manganese—75
Ni Nickel—75
Pb Lead—79
Pu Plutonium—81
Th Thorium—81
U Uranium—81
RE Lanthanum and Rare Earths (Lanthanides)—83
Ce Cerium—83
Dy Dysprosium—83
Er Erbium—83
Gd Gadolinium—83
Ho Holmium—83
La Lanthanum—83
Lu Lutetium—83
Nd Neodymium—83
Pm Promethium—83
Pr Praseodymium—83
Sm Samarium—83
Tb Terbium—83
Tm Thulium—83
Yb Ytterbium—83
Sn Tin—84
Ti Titanium—85
Zn Zinc—88

3. Preparation of Special Ceramics and Cermets (Ceramography)—91

Oxides—92
Carbides—94
Nitrides—97
Borides—98
Phosphides and Sulfides—98
Cermets—99
Iron Oxide Layers on Iron—100

APPENDIX A: Suggestions for Handling Hazardous Materials—103

APPENDIX B: Chemicals Used to Prepare Etchants in Chapters 2 and 3—105

APPENDIX C: References—108

Safety and Toxicology—108
General Textbooks and Reviews—109
Metallographic Preparation—110
Handbooks, Compilations, and Tables—111
Journals of Mostly Metallographic Content—112

APPENDIX D: Some Suppliers of Metallographic Equipment and Materials—113

North America—113
South America—114
Central America—116
Europe—116
Asia—121
Australia—124
New Zealand—124
Middle East—124
Africa—126

Index—129

Chapter 1: Technical Tips for Preparation of Metallographic Specimens

The methods of preparing metallographic sections for macroscopic and microscopic investigations are numerous and diverse. This is due to the variety of materials requiring investigation and the manner in which we have inherited much of the current data. Handed-down formulas have to be taken into consideration together with a more scientific approach to specimen preparation, patent rights, and commercial aspects as well as modern developments of improved preparation techniques and equipment. Thus, a comprehensive survey of specimen preparation is, at best, difficult. Nevertheless, some correlations are apparent and will be systematically explained prior to the tabulation of currently accepted metallographic etchants. There is no universal technique that will meet all the demands of metallographic specimen preparation.

Metallographic preparation normally requires a specific sequence of operations which includes sectioning, mounting, identification, grinding, polishing, cleaning, and etching. Each of these steps can be carried out in different ways and may vary according to the specific material properties. In principle, specimen preparation requires several steps, even though not all have to be pursued in every application. Care must be taken in performing each step because carelessness at any stage may affect later steps. In extreme cases, improper preparation may result in a false structure leading to erroneous interpretation.

A satisfactory metallographic specimen for macroscopic and/or microscopic investigation must include a representative plane area of the material. To clearly distinguish the structural details, this area must be free from changes caused by surface deformation, flowed material (smears), plucking (pull-out) and scratches. In certain cases, the edges of the specimen must be preserved. By observing simple common-sense principles, acceptable preparation is possible for any solid-state material, although in many cases it would require a lot of patience. Even for routine examinations in the least critical applications, poor specimen preparation is unacceptable because the observations and resulting conclusions are, at best, questionable.

Specimen Sectioning (Fig. 1)

The first step in specimen preparation—selection and separation of samples from the bulk material (sampling)—is of special importance. If the choice of a sample is not representative of the material, it cannot be corrected later. It is also difficult to compensate for improper sectioning, because additional, time-consuming corrective steps may be necessary to remove the damage produced.

Sectioning should render a plane surface suitable for further preparation without causing severe changes in the material. Alterations in the sample structure can be produced by deformation and the creation and further development of cracks and breakouts due to generation of heat, recrystallization, local tempering, and in extreme cases, partial melting. These problems can be minimized by the use of generous amounts of inert lubricants and coolants (water, oil, compressed air, etc.).

When sectioning with a torch or by normal mechanical sawing, cutting, sand blasting, or cleaving, care must be taken to cut sufficiently far from the area of interest to avoid harmful effects. The pre-sample must be large enough for the final sample to remain in the original state. This pre-sample may then be heavily ground or cut by more sophisticated and delicate means to produce the desired plane for further preparation. Ideally, only methods that produce surfaces suitable for immediate fine grinding or even polishing should be used. Such methods include abrasive cutting, ultrasonic chiseling, arc cutting, and electrochemical machining. Although the goal is to use such material-preserving methods for sectioning in the first place without the detour of a pre-sample, these methods

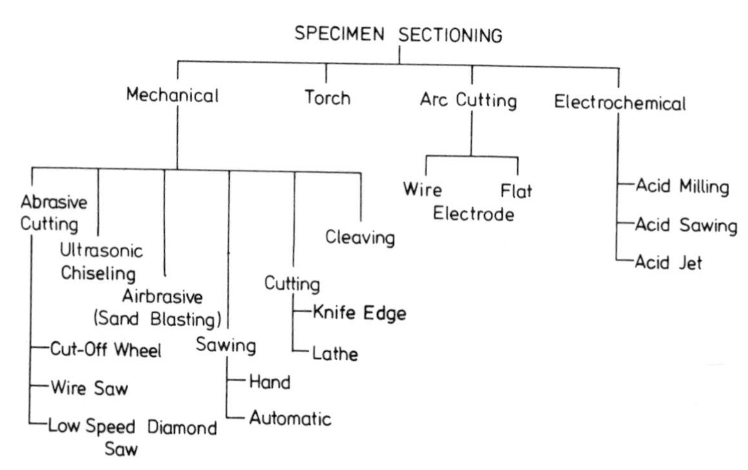

Fig. 1. Methods of metallographic specimen sectioning.

are troublesome and time-consuming and only useful for special applications (single crystals, semiconductors and brittle materials). They are not economical for routine work.

The most versatile and economical sectioning method is abrasive cutting. A thin, rapidly rotating, consumable abrasive wheel produces high-quality, low-distortion cuts in times ranging from 15 seconds to several minutes, depending upon the material and the cross-sectional area. This technique is almost universally applicable. Important parameters in abrasive cutting are wheel composition, coolant condition, and technique. Abrasive wheels consist of abrasive grains (alumina or silicon carbide) bonded together with either resin or rubber or a rubber-resin combination. By means of manufacturing controls, either "soft" (low wear resistance) or "hard" (high wear resistance) wheels may be produced. Diamond abrasive blades are either resin or metal bonded.

Generally, soft bonded cutoff wheels are used to cut hard materials; hard wheels are used for soft materials. The recent development of gravity-fed, low-speed (to 300 rpm) diamond sawing has provided a rapid and highly satisfactory means of sectioning small and delicate samples with minimal distortion.

Mounting (Fig. 2)

Mounting specimens in a holding device is necessary when preparing irregular, small, very soft, porous, or fragile specimens and in those cases where edge retention is required.

Embedding is indispensable when multiple specimens are to be included

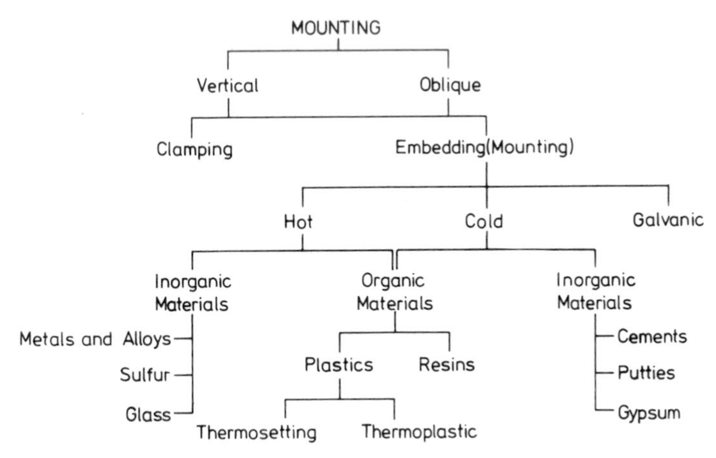

Fig. 2. Methods of metallographic mounting.

in a single mount or when automatic devices are to be used in further preparation. In most cases, mounting follows sectioning, but in the handling of a great number of very small specimens, it may be advantageous to reverse this order. In general, the mounting procedure can be easily adapted to the special problem in question. The existing multiplicity of mounting techniques may be reduced to those two basic methods generally used: clamping and embedding.

Clamping is a means of holding specimens (samples) together in a simple fixture device (Fig. 3). The specimens and the clamp should generally have similar mechanical and chemical characteristics so that neither polishing surface relief nor interference during etching occurs. Clamps are most commonly made of soft steels, stainless steel, aluminum alloys, or copper alloys. Stainless steel is the best choice for a wide variety of materials.

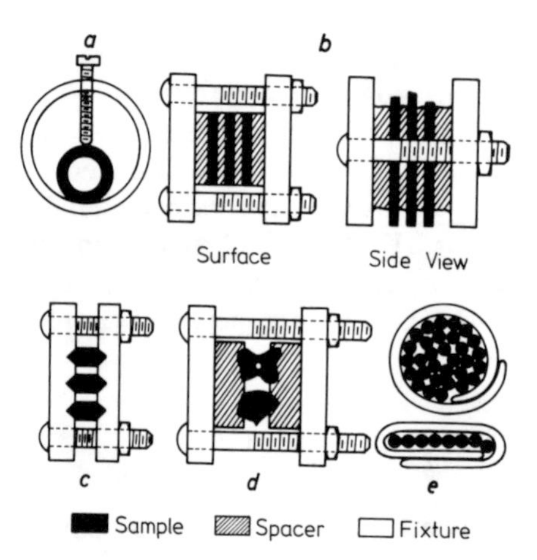

Fig. 3. Examples of fixtures used for clamping: (**a**) tube section, (**b**) sheet metal, (**c**) rods, (**d**) irregular pieces, (**e**) wires.

Shock-absorbing spacers (cork, rubber, or plastic) are sometimes used to reduce deformation but may interfere with preparation. A common problem encountered with clamps is the tendency of etching solutions and solvents to seep from gaps between the specimens and produce ugly stains.

Plastic embedding is the most popular method of mounting metallographic specimens at this time. The two main families of plastic used in metallographic mounting are the "hot" mounting and "cold" mounting resins:

Hot: Resin + Pressure + Heat = Polymer
Cold: Monomer + Catalyst (Heat) = Polymer + Heat.

Hot mounting (compression molding) requires the use of a press which exerts sufficient pressure (usually 29 MPa [4200 psi]) to a specimen mold heated to approximately 150 °C (300 °F). Under pressure and heat, the powdered resin fuses into a solid mass. Thermosetting resins, such as phenolics, cure and harden irreversibly with time; thermoplastics solidify when cooled to a lower temperature under pressure, but can be remelted.

Cold mounting (room-temperature-curing resins) requires uniform mixing of two components, usually powder and liquid or two liquids (a monomer and a catalyst). This mixture is cast into a suitable permanent, reusable mold (metal or glass ring on a glass plate) or a consumable mold (phenolic ring on a glass plate) which becomes part of the mount. The two chemical components react to form a solid mount (Fig. 4).

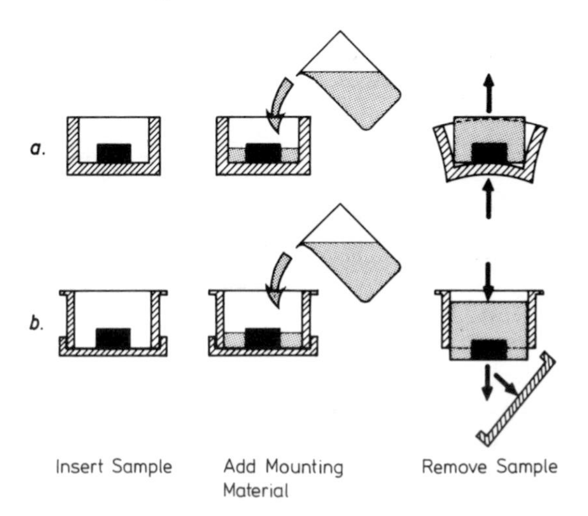

Fig. 4. Cold mounting of metallographic samples using (**a**) flexible, (**b**) inflexible, plastic molds.

Room-temperature-curing plastics are also advantageous when mounting specimens that are porous or contain fine cracks. With the aid of a vacuum chamber, pores and cracks may be filled and bubbles eliminated. This ***vacuum impregnation technique*** is particularly useful in the preparation of powder metal, ceramic, or porous specimens in general.

Thin layers, platings, or diffusion zones present a problem due to the difficulty in observing and measuring the thin areas. The apparent thickness of thin layers may be increased by means of a taper technique, in which the specimen is tilted at a shallow angle in the mold, as shown

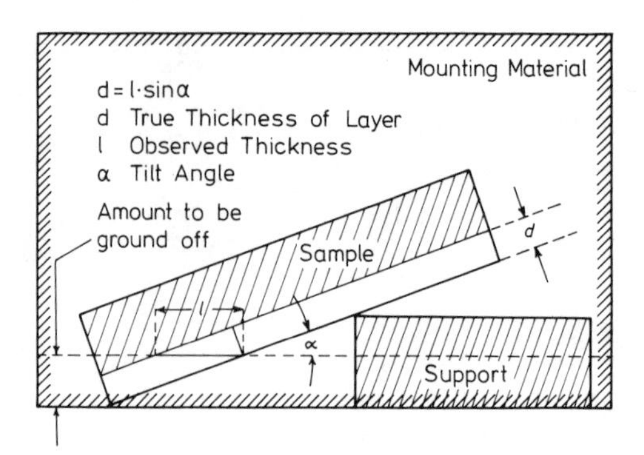

Fig. 5. Taper sectioning (oblique mounting to increase width of layer).

Table 1. Increase in Width of Layer as a Function of Tilt Angle (Compare Fig. 5)

Increase in layer width	Tilt angle	Increase in layer width	Tilt angle
25:1	2°20′	5:1	11°30′
20:1	2°50′	2:1	30°
15:1	3°50′	1.5:1	41°50′
10:1	5°40′		

in Fig. 5. The width advantage gained is dependent on the angle, as tabulated in Table 1.

Important properties of mounting materials (encapsulants) suitable for metallographic hot and cold mounting are:

- Inertness to sample material, etchant, mold, or any other reagent used in preparation
- Moderate viscosity in the liquid state and freedom from bubbles after solidification
- Low linear shrinkage
- Good adhesion to the specimen
- Hardness and wear resistance similar to those of the specimen

These requirements are satisfied by many of the commercially available thermosetting and thermoplastic mounting materials. While thermoplastics remain plastic in the characteristic temperature interval even after repeated heating, thermosetting materials harden irreversibly upon application of heat and pressure, and/or under the influence of a catalyst.

Although many mounting plastics possess the required characteristics, one type usually works best for a given application. Properties of encapsulants may be modified to make them suitable for special purposes. Metallic fillers such as silver or copper may be added to produce electroconductivity in electrolytic applications. Conductivity can also be achieved by drilling a hole through the mount from the rear up to the sample and inserting a conductor. Hardness or wear resistance may be increased through the addition of an abrasive filler such as alumina. This is particularly helpful in preserving the critical edges of hard specimens. Fillers are usually added to compression (hot) molding plastics during the manufacturing process. In room-temperature-curing encapsulants, the fillers are often added when mixing, prior to casting. Shrinkage gaps sometimes occur when the mounting plastic recedes from the sample surface during curing. A low-shrinkage epoxy resin or the preparation of critical surfaces with a pre-coating of plastic or electroplated metal may avoid or minimize this problem and allow mounting to proceed in the usual manner.

Heating of specimens, expected in compression molding, sometimes occurs in "cold" mounting also. Exothermic reactions in these encapsulants, particularly with acrylics, may reach as high as 150 °C (300 °F) during curing. This may also occur in epoxies and polyesters when the catalyst or hardener ratio is too high or if heat dissipation is poor. Mixing ratios should always be followed closely and cast mounts should be cured in a well-ventilated area, even a fume hood if available. Chilled water cooling may sometimes be required.

Materials which are temperature and/or pressure sensitive should be mounted in room-temperature-curing resins. Rigid materials that are not heat sensitive may be prepared more economically by compression molding. If only a few specimens are to be prepared, hot mounting is less time-consuming than cold mounting; with a larger number of specimens, the reverse holds.

Identification (Marking) (Fig. 6)

Specimens are useful only if their origin is known. Therefore, specimens should be clearly marked so that their identity is not lost at any time in the preparation sequence. It is important that marking does not cause any changes to the surface of the specimen. Methods such as stamping or vibration-engraving produce local deformation. In electroengraving, which utilizes the principle of spark evaporation, local changes in composition can take place through reaction with the surrounding atmosphere. These changes can alter the microstructure, through decarburization or remelting and quench cooling. Heat treated specimens are particularly susceptible to such effects. Markings should be placed on

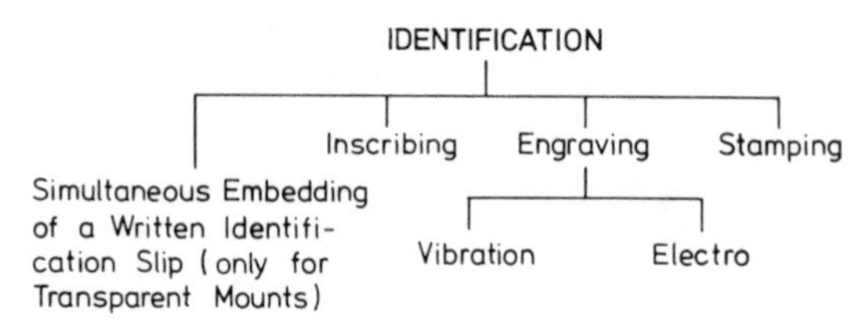

Fig. 6. Methods for identifying metallographic specimens.

the side opposite the critical face to avoid damage. Identification labels should be fixed on the mounts immediately to avoid loss of identity at this point.

Grinding and Polishing (Fig. 7)

Coarse and fine grinding are the initial steps in producing a satisfactory polished specimen. Together with polishing, grinding serves to smooth the surface, remove the altered surface material, and reveal the true structure of the specimen. These processes are carried out in several steps from coarse to fine so that the deformed layer may be effectively reduced. Coarse grinding removes heavy surface damage and produces a flat surface comparable to filing or planing, which are seldom used in metallography. The surface roughness resulting from coarse grinding is in the range of 10 to 100 μm (1 μm = 10^{-3} mm = 3.9×10^{-5} in.). Grinding leaves a roughness of 1 to 10 μm, and fine grinding of about 1 μm. Although grinding produces a flat, smooth surface, it is suitable for macroscopic examination or macroetching only.

Because there is not a sharp line between grinding and polishing,

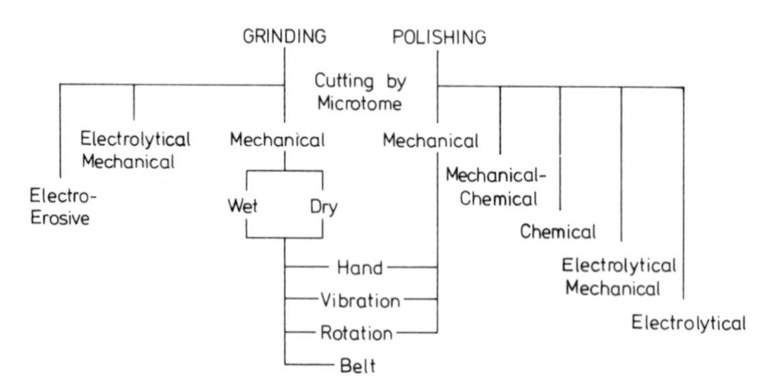

Fig. 7. Methods of metallographic grinding and polishing.

there is considerable overlap between fine grinding and coarse polishing. For this reason, grinding and polishing are described here together.

Mechanical Grinding and Polishing

Mechanical grinding and polishing are most common. They are usually accomplished by abrasive removal, using a manual technique on fixed abrasives, vibrating platens, rotating wheels, or continuous belts.

During mechanical grinding and polishing, the surface is removed by the abrasive in the form of shavings or chips. To a lesser degree, removal takes place through adhesive interaction between sample and polishing support. Scratches, deformation, and some smearing are characteristic consequences of mechanical grinding and polishing. The purpose of grinding and polishing is to produce a suitable surface by means of a step-by-step removal of deformed material until it becomes negligible.

While deformation is predominant in coarse grinding, smearing due to plastic flow is more prevalent during polishing. This is illustrated schematically in Fig. 8. It is assumed that polishing produces a flat

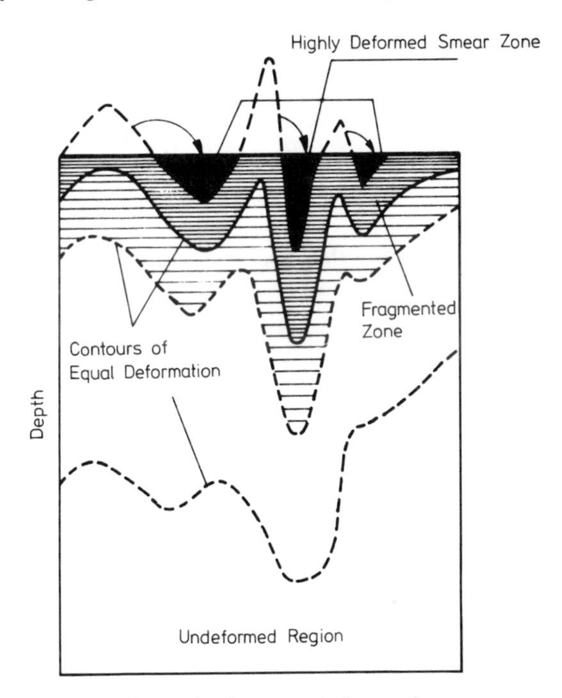

Fig. 8. Schematic diagram of the surface region of a microsection after grinding and polishing. (Section at right angle to surface.)

surface and that grinding scratches have been removed. The scratch troughs have been filled with material previously protruding above the surface. This is the result of the rolling action of loose, unsupported abrasive particles in the cloth polishing steps. This filling effect gives rise to the smeared black layer in Fig. 8. The black region is composed of severely deformed material from the specimen, mixed with polishing compound. Adjacent to this layer is a range of deformed material with a decreasing severity of deformation toward the interior of the specimen. Lines of equal deformation do not run parallel to the specimen surface, but follow the profile generated by the coarse grinding action.

Varying surface conditions produce different properties such as electrochemical potentials. Areas having the highest degree of deformation, for example, have the greatest reaction to etching solutions and are, therefore, dissolved at a higher rate. This is the reason for the reappearance of scratches on polished surfaces after etching, revealing poor preparation.

The depth of scratches (roughness) plus the depth of deformation equals the total depth of disturbed material as shown in Fig. 9. This is based on references from literature using steel as the example. The depth of roughness is directly proportional to the particle size of the abrasive, but the depth of deformation approaches a constant value after the initial increase.

$$\text{Depth of Roughness} + \text{Depth of Deformation} = \text{Depth of Disturbed Material}$$

An understanding of the relationship illustrated in Fig. 9 is necessary

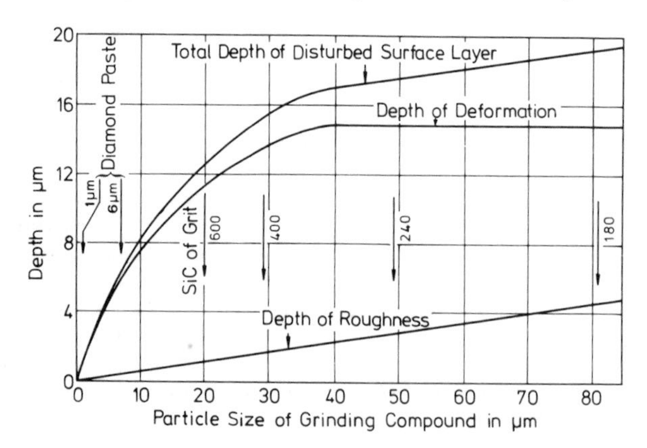

Fig. 9. Depth of roughness, deformation, and total depth of disturbed surface layer as a function of particle size of grinding and polishing compounds.

for the successful application of abrasive polishing. In practice, the curves are followed from right to left or from coarser to finer abrasive sizes. Removal of all deformation is the key to successful optical measurements and microhardness testing.

It is not adequate to remove roughness (visible scratches) from the surface without removing the deformed layer as well. Coarse grinding should be kept to a minimum because there is a tendency to produce severe deformation. Since depth of deformation decreases sharply after a particle size of approximately 30 μm, which corresponds to a grit size of about 400, increased time should be spent in removing previous damage. Care must be taken to prevent cross contamination of coarse abrasive particles in the finer steps (careful cleaning between steps!).

A number of parameters are influential during mechanical grinding and polishing. These will be explained subsequently.

Grinding and Polishing Compounds. Inorganic powders are used as abrasives in metallography. They are listed in Table 2 in order of increasing Mohs hardness from pumice (the softest) at the top to diamond (the hardest) at the bottom. This indicates that a given material is able to scratch the one above it on the scale. Also shown is Rosiwal's hardness scale using corundum at 1000 as a reference value.

Figure 10 illustrates the common areas of application for the more usual grinding and polishing abrasives. In principle, any one abrasive

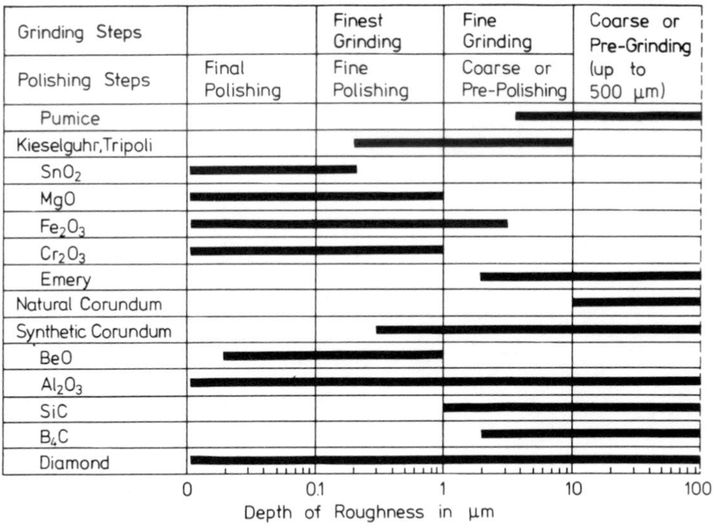

Fig. 10. Ranges of application of metallographic grinding and polishing compounds.

Table 2. Abrasives and Polishing Compounds

Standard minerals of Mohs scale	Rosiwal grinding hardness	Materials for comparison	Abrasives and polishing compounds	Remarks
		Graphite		
1 Talcum	0.03	Lead, tin		
2 Gypsum, rock salt	1.25	Aluminum, zinc, magnesium, copper, silver, gold		
3 Calcite	4.5	Marble, brass, iron, nickel		
4 Fluorspar	5	Steels, unhardened		
5 Apatite	6.5	Window glass	**Pumice** (mixtures of SiO_2 and Al_2O_3).	Mild abrasive. Occasionally for silver.
			Kieselguhr and **tripoli** (silicic acid and alumina mixtures).	Prepolishing paste for brass and precious metals.
			Tin dioxide (SnO_2).	For fine polishing of ores.
6 Feldspar	37	Low-carbon steels, hardened	**Magnesia** (MgO).	For polishing of magnesium and aluminum and their alloys.
			Cerium oxide (Ce_2O_3)	Rarely used in metallography. Mostly for glass.
			Iron oxide (Fe_2O_3; polishing red).	For polishing of soft metals and alloys.
7 Quartz	120	0.9% C steel, hardened; tungsten	**Quartz** (SiO_2, sand).	Not used in metallography.
8 Topaz	175	Special steels	**Chromium oxide** (Cr_2O_3).	For polishing of hard metals and alloys; for example, chromium, steels.
			Emery (50–70% Al_2O_3; remainder SiO_2, Fe_2O_3 and Fe_3O_4).	Seldom used abrasive and polishing compound.

Table 2. Abrasives and Polishing Compounds (*continued*)

Standard minerals of Mohs scale	Rosiwal grinding hardness	Materials for comparison	Abrasives and polishing compounds	Remarks
9 Corundum	1000			
			Natural corundum (90–95% Al_2O_3).	Abrasive wheels for steels, brasses, etc.
			Synthetic corundum or **electrocorundum** (60–95% Al_2O_3; remainder SiO_2, TiO_2, Fe_2O_3; variations: precious corundum, alundum, and corundum with $Al_2O_3 > 99\%$).	Important abrasive for steel, cast steel, sintered metals, hard steels.
			Beryllium oxide (BeO).	Toxic! For polishing of cemented carbides.
			Alumina (Al_2O_3; modification: cubic γ-alumina, hexagonal α-alumina).	Commonly used polishing compound. Universally suitable for metallographic purposes.
			Silicon carbide (SiC; carborundum).	Commonly used abrasive. Universally suitable for metallographic purposes.
			Boron carbide (B_4C).	For grinding of cemented carbides.
10 Diamond	14000		**Diamond.**	For grinding and polishing of most metals, glasses, precious stones.

could be used over the entire range of specimen preparation—a real possibility with diamond and alumina. Other abrasives are not as versatile, due to economics, availability, and efficiency of application. The subdivision into several grinding and polishing steps shown in Fig. 10 is arbitrary and not intended to be strictly observed; it is presented as a useful guide in practice.

Uniform particle size of known values is essential for good metallographic abrasives. Different size designations are in use. Figure 11 shows the relationship between mesh or grit size, emery grade, and true particle size in microns. Mesh is a number used to denote the size of individual

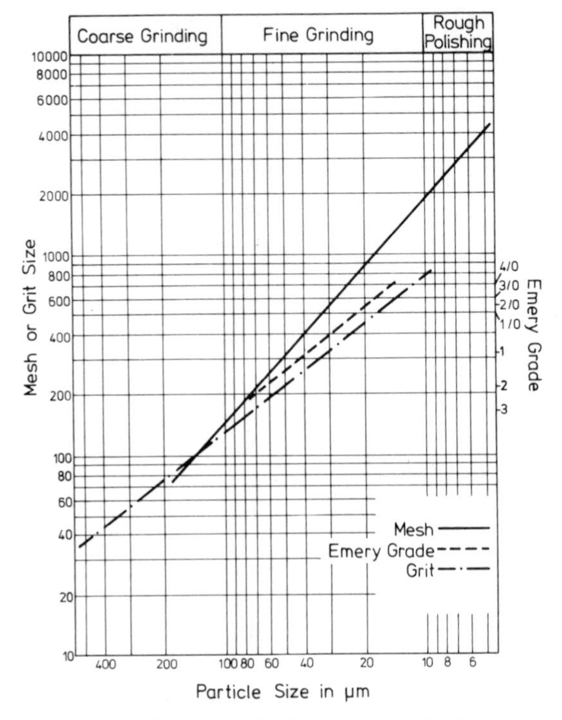

Fig. 11. Relationship between mesh size, grit size, emery grade, and particle size in microns.

abrasive grains. This corresponds to the screen openings per linear inch in the standard sieve. For example, an abrasive of 320 mesh contains particles that will just pass through a screen having 240 openings to the linear inch, but will be retained by the next-finer, 320-mesh screen. In addition to mesh size, grit size and emery grade are also commonly used as standards to classify abrasive materials.

Sharp edges, high hardness, high coating density, and good bonding to the support material substantially increase the cutting rate and reduce the depth of deformation. Diamond is superior to the other abrasives because of the degree to which it meets the above requirements.

Grinding and Polishing Fluids. Although some specimen-preparation work is done in air or inert gas without liquid vehicles, this is the exception. Normally, all metallographic preparation steps require a liquid vehicle as a coolant and/or dispersant, causing loose abrasives to be distributed more uniformly onto cloth surfaces. Wet polishing has numerous advantages, which include:

- Control of heat at the specimen-abrasive interface.
- Control of harmful dust.
- Longer life for fixed abrasives because removed products are continuously flushed away.

Pressure, Time, Velocity. Increasing the value of pressure, time, and velocity generally works toward a higher material-removal rate. The effect on deformation varies such that:

- Excessive pressure may cause heating and flowed material, which can cause changes in microstructure. Pressure should not be too high, especially during polishing.
- Short grinding times and long polishing times are preferable (Fig. 9). However, cloth polishing with abrasives other than diamond should be as brief as possible, because severe relief effects may result from the differing removal rates for individual microstructural components.
- The harder the specimen material, generally the lower the applied grinding and polishing speed. However this "rule" is only of limited validity. For some extremely hard materials (ceramics, intermetallic compounds, cemented carbides), higher polishing speeds are preferred in most cases.
- Because increased grinding and polishing speeds produce higher surface temperatures, heat-sensitive materials must be polished at lower speeds.

Specimen Motion in Grinding and Polishing. The motion of the specimen during grinding and polishing operations affects edge retention. For the best results, the specimen must be held flat against the abrasive surface at all times. To avoid the formation of oriented grinding and polishing grooves, these operations are best performed by rotating the specimen 90° between each step. When using a wheel for grinding and polishing, rotating the specimen opposite to the wheel rotation eliminates directional effects.

Grinding and Polishing Substrates. A wide variety of substrate materials are used in metallographic specimen preparation. Paper, cloth, metal, wood, glass, hard rubber, and pitch have all been used as supports in abrasive preparation. Coarse grinding is most commonly performed on coated paper discs or cloth belts. Fine grinding is usually executed on coated paper abrasives; rough and final polishing is almost always done on fabrics such as wool, silk, cotton, felt, and various synthetic materials. The choice of cloth type is very important; certain cloths work well in certain applications but not in others. Low-nap cloths are generally preferred for rough polishing and medium-nap cloths for final polishing. Abrasives are usually applied to cloths in the form of paste suspensions or slurries.

- Hard or firm support materials tend to promote deeper deformation.
- More elastic support materials favor less deformation but increase the tendency to relief and edge rounding.

Microtome Cutting

Microtome cutting employs a cemented carbide or diamond knife which mechanically slices a layer from the bulk material. Although the action of a microtome is similar to planing or milling, the principles of mechanical grinding and polishing still apply. High-quality cuts, free from deformation and flow, are possible with the microtome only if the cutting angle is properly adjusted. This technique is particularly useful in cutting aggregate materials such as lamellar structures, having varying degrees of hardness. Microtome cutting is, however, restricted to materials of about 150 HV or less.

Similar to microtome cutting is micromilling. Whereas the knife of a microtome removes a surface layer by cutting and acts like a plane, the micromill has a rotating milling head which removes layers 5 to 15 μm in thickness with high precision. As with the cutting knife of the microtome, the milling tool of the milling device is made of either diamond or cemented carbide. Micromilling combines all grinding and polishing operations into one step and produces a plane surface of high quality with respect to scratches and surface damage.

Electrolytic Grinding and Polishing

Although electrolytic grinding and electroerosive grinding (Fig. 7) are seldom used, electrolytic polishing is commonly employed. Electrolytic polishing, also called anodic polishing, occurs through anodic dissolution of the specimen surface in an electrolytic cell.

Figure 12 is a diagram showing a simple electrolytic cell that is easily set up in the laboratory. The orientation of the anode (sample) to the cathode may be adapted to suit the particular application. Commercially designed and produced instruments are available which are versatile enough to meet most laboratory requirements.

Electrolytes suitable for metallographic purposes are usually mixtures of acids such as phosphoric, sulfuric, and perchloric in ionizing solutions such as water, acetic acid, or alcohol. Glycerol, butyl glycol, urea, etc. are added to increase the viscosity. Metals which form highly soluble

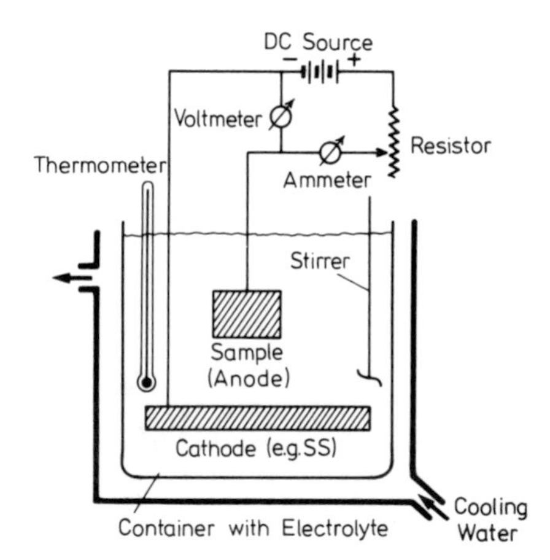

Fig. 12. Electrolytic polishing cell in series mode.

hydroxides are prepared with alkaline solutions, while those forming highly soluble cyanides are treated in cyanides. Most of the electrolytes mentioned in Chapters 2 and 3 are harmless when handled according to known common-sense precautions. Mixtures of perchloric acid, however, are particularly prone to decompose violently and should, therefore, be treated with extreme care. (See Appendix A.)

The tendency of perchloric acid mixtures to explode is related to the concentration. This is illustrated in Fig. 13, the ternary diagram for perchloric acid, water, and acetic acid. Mixtures outside the immediate danger zone can undergo local changes in concentration by means of evaporation, temperature increase, contact with organic material, or contact with bismuth or sparks. If this occurs, the solution may move into the danger zone without the operator's awareness. To minimize such dangers, the electrolyte should be stirred and cooled. In any case, and at all times, caution is advised when using perchloric acid.

Significant parameters which affect the results of electropolishing are:

- Current density (A/cm^2)
- Voltage (V)
- Electrolyte composition, temperature, and flow rate
- Polishing time
- Initial condition of the specimen surface
- Cathode size, shape, and composition.

Most formulas in Chapters 2 and 3 specify these details; they are omitted when they are not critical or can easily be established by the operator.

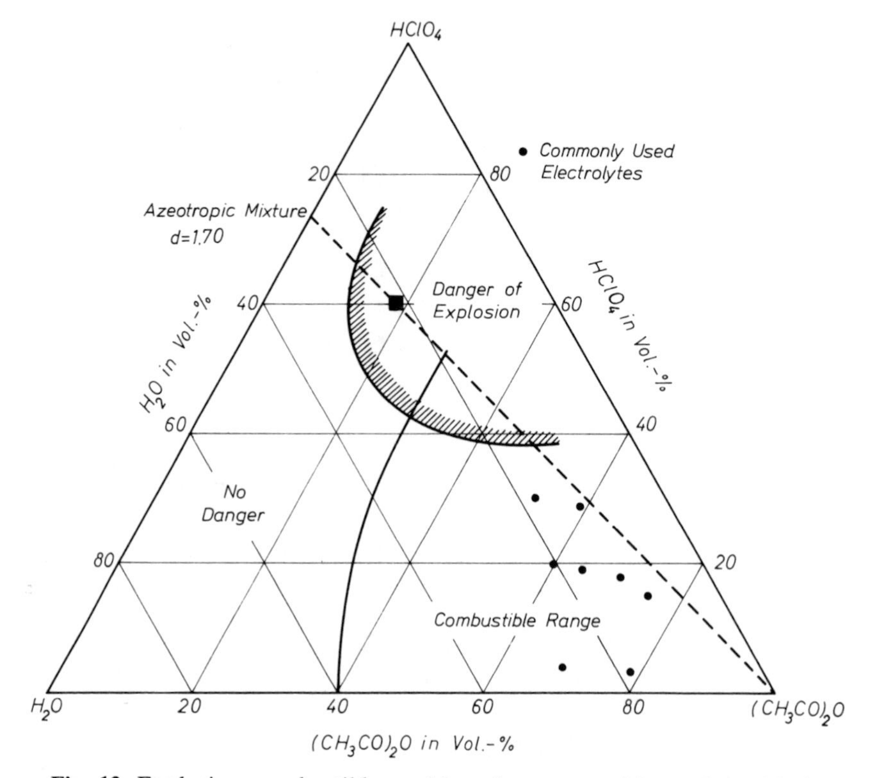

Fig. 13. Explosive, combustible, and harmless compositions of perchloric acid – acetic anhydride – water mixtures.

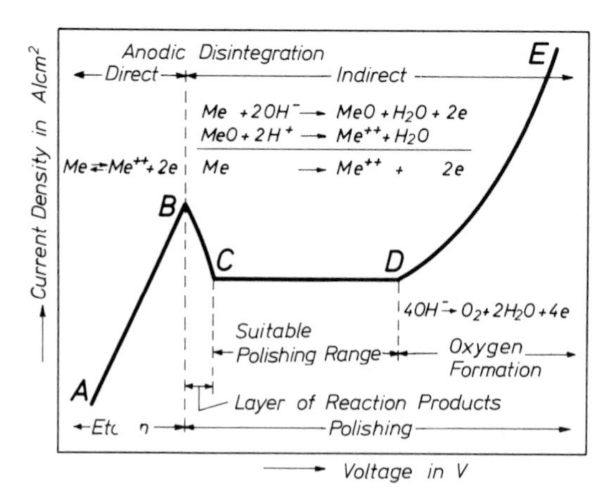

Fig. 14. Idealized current density versus applied voltage. See text for discussion.

A typical idealized current density versus applied voltage relationship for many common electrolytes is shown in Fig. 14, with four characteristic regions noted:

1. *A to B.* The anode material goes directly into solution. A liquid layer forms at the surface with a higher metal-ion concentration than is present in the rest of the solution. Electrolytic etching occurs in this area.
2. *B to C.* It is assumed that formation of a thin layer of reaction products causes passivity.
3. *C to D.* Polishing proceeds due to diffusion and electrochemical processes.
4. *D to E.* Oxygen evolution occurs at a low rate, initially. Gas bubbles adhere to the surface of the anode and remain there for a relatively long time, with pitting as a result. With increasing voltage, the rate of oxygen evolution increases and the bubbles remain for shorter periods of time, until they do not adhere at all.

Although anodic polishing begins at point B of the curve, only between points C and D is there freedom from other negative effects (passivation, oxygen formation). Metallographic specimen preparation, therefore, is concerned with area C-D of the current density versus voltage curve. Region D-E is rarely used and then only in industrial applications. The different regions illustrated in Fig. 14 are not always clearly observed. Electrolytes of high specific resistance produce rather flat curves.

The initial condition of the specimen surface has a definite influence on the polishing time. In general, the better the surface at the start and the higher the current density, the shorter the polishing time.

Electrolytic polishing requires electrical conductivity of the sample material; this is the case for all metals and alloys, as well as for some nonmetals, such as carbides and graphite. In principle, all homogeneous metals and alloys are suitable for electrolytic preparation; heterogeneous materials such as gray iron are not ideally suited for electropolishing due to electrochemical potential differences between the various phases. However, examples of electropolishing of heterogeneous alloys are becoming increasingly common. Some advantages of electropolishing are:

- Freedom from deformation and flowed material
- Rapidity and reproducibility. However, this is only valid if suitable conditions are established
- No (or minimal) surface heating
- Capability of sequential electroetching with the same instrument

- Possible removal of mechanical damage caused by mechanical grinding and polishing.

Restrictions on the use of electropolishing are imposed by the lack of etchants for many heterogeneous alloys as well as by the following disadvantages:

- Edges are selectively attacked, producing a radius and resulting in poor edge preservation.
- Macroscopic out-of-flatness cannot be remedied.
- Residues may be deposited on the surface during polishing.
- Oxidation of some specimens may interrupt the process.
- If the mounting material is nonconductive, special mounting procedures are necessary to produce electroconductivity.
- Coarse-grain materials are less suited for electropolishing.
- Depressions may occur around inclusions because of the higher solution rates of the metallic material in that zone.

Chemical Polishing

Chemical polishing is a process by which simple immersion of a specimen in a suitable polishing solution (electrolyte) produces a polished surface without the use of externally applied current. When the specimen is agitated in the polishing bath (seconds to minutes), surface roughness is removed and a deformation-free polished surface is produced.

The polishing solutions almost always contain oxidizing agents such as nitric, sulfuric, and chromic acids or hydrogen peroxide. Viscous agents are also added to control diffusion and convection rates, producing a more uniform process. Phosphoric acid, for example, forms a liquid film of high viscosity when reacting with metal ions. Most chemical polishing solutions are quite insensitive to concentration changes.

The advantages of chemical polishing are:

- Simplicity and economy
- Little pre-preparation required; 320-grit finish is adequate; specimens cut with an abrasive cutoff wheel may be polished without further preparation
- Simple treatment after polishing; in most cases, rinsing in water is adequate
- No deformation or flow lines produced
- Specimen and/or mount need not be electrical conductors.

Disadvantages of chemical polishing are:

- Higher rate of attack at edges

- Orange-peel effect with coarse-grain materials
- Only small-scale roughness may be smoothed out
- Formation of a surface film composed of reaction products.

In spite of many advantages, chemical polishing is seldom used for metallographic preparation. This may be due to a lack of awareness by practicing metallographers, and by the limited number of published formulas.

Combination Polishing Methods

When a single method of specimen preparation fails to produce the desired results, a combination of methods may work very well. Some examples are described below.

Etch-Polishing Sequence. Flowed metal layers may be effectively removed by lightly etching and mechanical repolishing one or more times at the conclusion of final mechanical polishing.

Attack-Polishing. This consists of the simultaneous application of a chemical etchant during the final polishing sequence; it is especially useful in avoiding flow layers on soft metals and alloys as well as refractory metals.

Multiple Polishing. When electrolytic polishing is used with heterogeneous alloys, some undesirable surface effects like relief formation or surface layers may be produced. A brief intermediate mechanical polishing step may remove this condition and produce a satisfactory finish.

Electrolytic Lapping. This is the simultaneous electrolytic and mechanical removal of material. In this technique, the specimen (anode) is held against a cloth saturated with electrolyte and mounted on a support or wheel which serves as the cathode. This method, illustrated in Fig. 15, is considered semiautomatic. Although direct current is most frequently used in electrolytic lapping, low-frequency alternating current produces superior results with some metals, such as molybdenum,

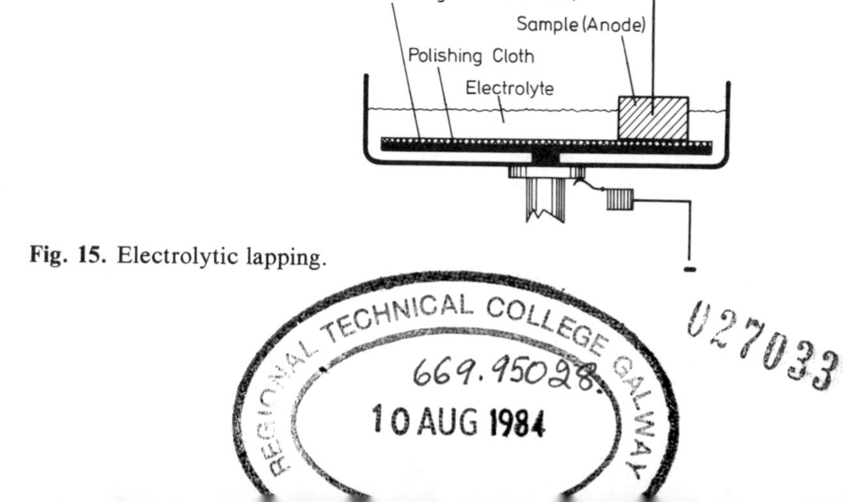

Fig. 15. Electrolytic lapping.

tungsten, and rhenium. Dilute solutions of sodium thiosulfate, nitric acid, oxalic acid, picric acid, and hydrogen peroxide are suitable for both electrolytic lapping and etching. The addition of abrasives such as alumina has no effect in most applications.

Electrolytic lapping combines the advantages and disadvantages of electrochemical and mechanical polishing. It can be used for heterogeneous as well as homogeneous metals and alloys, and has proven successful as an addition to other methods; in some cases, it succeeds when electrolytic or mechanical polishing alone fails.

Automatic Grinding and Polishing

Manual specimen preparation tends to be routine, monotonous, and even tedious. This is especially true of mechanical polishing. When operators become tired and disinterested, the work they produce may decrease in quality and volume. Preparation errors are more likely to occur, which can ultimately result in erroneous interpretations. Thus, it is understandable that again and again attempts have been made to automate the mechanical fine grinding and polishing steps.

Prior to 1950, manual preparation was the only means considered satisfactory for the sensitive grinding and polishing steps. Since then, there has been a considerable upswing in the development of precision instruments for automation of metallographic preparation. For the most part, these have been confined to individual steps; and, to this date, no satisfactory fully automatic system has been developed. Specimens must still be handled between steps, but the ideal system would be to insert samples at one end and remove completely polished specimens at the other end. This would be particularly helpful in preparing materials requiring numerous lengthy steps. Nevertheless, the automation of the single steps increases quality, volume, and uniformity by removing most of the physical effort from specimen preparation.

Evaluation of Polishing Methods

Reflectivity of the finish-polished surface is one sensitive criterion for judging quality. Deviations from the reflectivity of ideal surfaces with an atomic roughness give an indication of the effect of different polishing methods; for example, cleavage surfaces of crystals or surfaces of single crystals grown under an ultrahigh vacuum are good reference standards. Therefore, reflectivity is suitable in evaluating the effectiveness of individual polishing methods.

Figure 16 shows the mean deviation, in percentage of reflectivity, of polished surfaces from ideal surfaces of various materials. The base line represents reflectivity values for freshly cleaved surfaces—the ideal

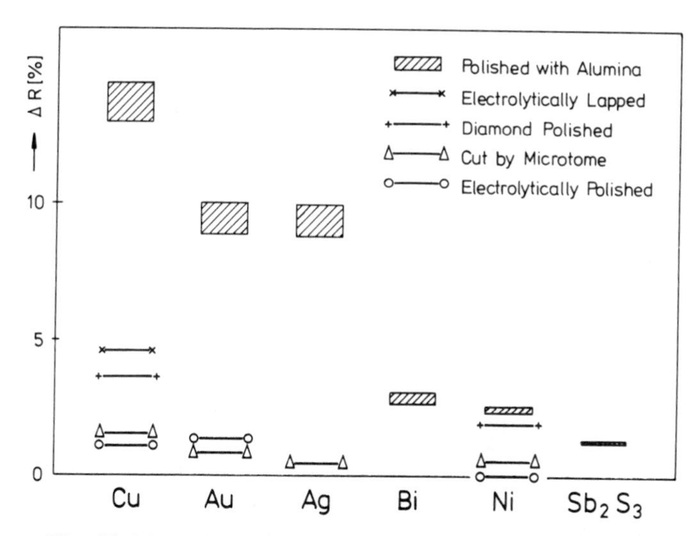

Fig. 16. Mean deviations ΔR, in percentage of reflectivity, of a polished surface from an ideal surface.

values of surface quality obtainable. Mean deviations for different polishing methods are plotted for different materials.

Obviously, any polishing treatment will produce a lower quality than the ideal cleaved surface. Of the popular polishing techniques, alumina polishing produces the poorest values while electrolytic and diamond-knife microtoming will produce results approaching the ideal values.

It is also obvious from Fig. 16 that the effectiveness of different polishing methods varies from material to material. Antimony trioxide (Sb_2O_3), for example, shows the same deviation when prepared with alumina slurries as gold and copper do after electrolytic polishing. Properties other than reflectivity are also affected by the polishing technique. Severe grinding and polishing stresses have a noticeable affect on microhardness. When reporting surface condition, information concerning the grinding and polishing techniques should be included. Differences in actual results compared with expected results may stem from differences in the manner of preparation. For routine preparation and examination of microsections, these considerations are of secondary importance. Alumina and other oxides are widely used because they produce acceptable results at a reasonable cost.

Cleaning (Fig. 17)

Cleanliness is an important requirement for successful metallography. Specimens must be cleaned after each step; all grains from one grinding

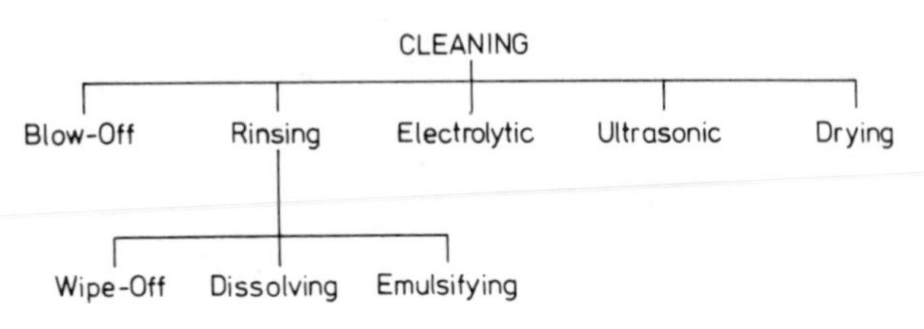

Fig. 17. Methods of cleaning metallographic specimens.

and polishing step must be completely removed from the specimen to avoid contamination, which would reduce the efficiency of the next preparation step. Thorough cleaning is particularly critical after fine grinding and before rough polishing and all subsequent steps.

Clean, grease-free surfaces are essential for subsequent chemical or electrolytic treatment. Residues, fingerprints, and inconspicuous films may interfere with etching, causing various areas to be attacked at different rates. Every single microsection-preparation procedure must be followed by thorough cleaning, which can be performed in different ways.

Rinsing is most frequently used and consists of holding the specimen under a stream of running water and wiping the surface with a soft brush or cotton swab.

Ultrasonic cleaning is the most effective and thorough method of cleaning. Not only are surface contaminants removed, but particulate matter held in crevices, cracks, or pores is removed by the action of cavitation. Usually this ultrasonic cleaning needs only 10 to 30 s.

After cleaning, specimens may be dried rapidly by rinsing in alcohol, benzene, or other low-boiling-point liquids, then placed under a hot-air drier for sufficient time to vaporize liquids remaining in cracks and pores.

Etching (Fig. 18)

Although certain information may be obtained from as-polished specimens, the microstructure is usually visible only after etching. Only features which exhibit a significant difference in reflectivity (10% or greater) can be viewed without etching. This is true of microstructural features with strong color differences or with large differences in hardness causing relief formation. Cracks, pores, pits, and nonmetallic inclusions may be observed in the as-polished condition.

In most cases, a polished specimen will not exhibit its microstructure because incident light is uniformly reflected. Since small differences

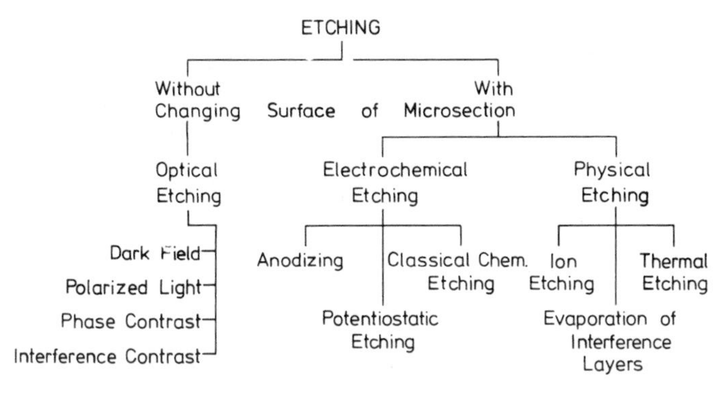

Fig. 18. Methods of metallographic etching.

in reflectivity cannot be recognized by the human eye, some means of producing image contrast must be employed. Although this has become known as "etching" in metallography, it does not always refer to selective chemical dissolution of various structural features.

There are numerous ways of achieving contrast. These methods may be classified as "optical," "electrochemical (chemical)," or "physical," depending on whether the process alters the surface or leaves it intact.

Optical Etching

"Optical etching" is based on the application of certain illumination methods, all of which use the "Köhler" illumination principle. This principle also underlies common bright-field illumination. These illumination modes are dark field, polarized light, phase contrast, and interference contrast. They are available in many commercially produced microscopes, and in most cases, the mode may be put into operation with a few simple manipulations; in other cases, it involves the addition of accessories. Image quality is not reduced significantly, although the intensity of the light at the image plane is affected considerably by using different illumination modes. These illumination possibilities are not yet as fully exploited as one would assume, judging from the simplicity of manipulation and the dependability of the results. There is a distinct advantage in employing optical etching rather than those techniques which alter the specimen surface. Chemical and physical etching require considerable time and effort and there is always a danger of producing artifacts which lead to misinterpretations.

Electrochemical (Chemical) Etching

During the process of electrochemical etching (often termed "chemical" etching) of metallic specimens, reduction and oxidation processes (redox processes)* take place. All metals in contact with the solutions have a pronounced tendency to become ionized by releasing (losing) electrons. The degree to which this reaction takes place may be recorded by measuring the electrochemical potential. This is done by comparing the potential of metal versus the standard potential of a reference electrode.† The tabulation of various metals results in the electromotive series of elements:

Li^+, Na^+, K^+, Ca^{++}, Ba^{++}, Be^{++}, Mg^{++}, Al^{+++}, Mn^{++}, Zn^{++}, Cr^{+++}, Cd^{++}, Tl^+, Co^{++}, Ni^{++}, Pb^{++}, Fe^{+++}, H^+, Sn^{++++}, Sb^{+++}, Bi^{+++}, As^{+++}, Cu^{++}, Ag^+, Hg^{++}, Au^{+++}, Pt^{+++}.

The elements are listed in order of decreasing electroaffinity. All elements preceding hydrogen are attacked by acids with the evolution of hydrogen (H_2). All elements following hydrogen cannot be attacked without the addition of an oxidizing agent. Thus, microstructural elements of different electrochemical potential are attacked at different rates. This produces differential etching, resulting in microstructural contrast.

Electrochemical etching may be considered as "forced corrosion." The differences in potential of the microstructural elements cause a subdivision into a network of very small anodic or cathodic regions (local elements). These miniature cells cannot originate from differences in phase composition only, but also have to come from irregularities in the crystal structure as they are present—for example, at grain boundaries and from other inhomogeneities such as:

- Inhomogeneities resulting from deformation (deformed zones), which are less resistant to attack than undeformed material.
- Unevenness in the formation of oxidation layers (glossy areas are less resistant).
- Concentration fluctuation in the electrolyte (low concentration is less resistant).
- Differences in electrolyte velocity (higher circulation rates reduce resistance to attack).
- Differences in the oxygen content of the electrolyte (aerated solutions are more resistant).

---

* Reduction = absorption of electrons (cathodic reaction)
Oxidation = emission of electrons (anodic reaction)

Oxidizing agent + electron ⇌ reducing agent
Fe^{3+} + e ⇌ Fe^{2+}

Electrolytic grinding and polishing, chemical polishing, and some of the combined methods are based on redox processes. This is not always obvious from conventional terminology.

†Hydrogen electrode: a sheet of platinum surrounded by hydrogen at 1 atm pressure and immersed into an aqueous solution of hydrogen-ion activity $a_H = 1$.

- Differences in the illumination intensity, which can initiate differences in potential.

Because of differences in potential between microstructural features, dissolution of the surface proceeds at different rates, producing contrast. Contrast can also originate from layers formed simultaneously with material dissolution. This is true in precipitation etching and heat tinting where surface reactions are involved. In *precipitation (deposit) etching*, the material is first dissolved at the surface; it then reacts with certain components of the etchant to form insoluble compounds. These compounds precipitate selectively on the surface, causing interference colors or heavy layers of a specific color. During *heat tinting*, coloration of the surface takes place at different rates according to the reaction characteristics of different microstructural elements under the given conditions of atmosphere and temperature.

A wide variety of etchants is available, including acids, bases, neutral solutions, mixtures of solutions, molten salts and gases; many examples of these are given in Chapters 2 and 3. Most of these formulas were derived empirically. Their composition and mode of application can be easily varied and modified, and so also be useful for materials other than those mentioned in the formulas. The rate of attack is mainly determined by the degree of dissociation of the etchant and its electrical conductivity. Both are often influenced in a certain way by small additions of other chemicals. This may explain why many formulas contain small amounts of substances whose significance is not immediately apparent.

The stability of many etching solutions is limited; redox potentials change with time. Changes may also occur while the etchant is in use, such that it must be discarded after a limited time.

Etching times range from several seconds to some hours. When no instructions are given, progress is judged by the appearance of the surface during etching. Usually, the surface will become less reflective (duller) as etching proceeds. Etching time and temperature are closely related; by increasing the temperature, the time can usually be decreased. However, this may not be advisable because the contrast could become uneven when the rate of attack is too rapid. Most etching is performed at room temperature.

Sources of error are numerous, especially in electrochemical etching. Etching errors may lead to microstructural misinterpretation. For example, precipitates from etching or washing solutions could be interpreted as additional phases.

Conventional chemical etching is the oldest and most commonly applied technique for producing microstructural contrast. In this technique, the etchant reacts with the specimen surface without the use of an external current supply. Etching proceeds by selective dissolution according to the electrochemical characteristic of the component areas.

In *electrolytic* or *anodic etching,* an electrical potential is applied to the specimen by means of an external circuit. Figure 12 illustrates a typical setup consisting of the specimen (anode) and its counterelectrode (cathode) immersed in an electrolyte (etchant). During anodic etching, positive metal ions leave the specimen surface and diffuse into the electrolyte with an equivalent number of electrons remaining in the material. This results in a direct etching process shown as segment A-B of the current density versus voltage curve in Fig. 14. Sample dissolution of material without the formation of a layer occurs in this instance. If, however, the metal ions leaving the material react with nonmetal ions from the electrolyte with formation of an insoluble compound, precipitated layers of varying thickness will form on the specimen surface. The thickness of these layers is a function of the composition and orientation of the microstructural features exposed to the solution. These layers may reveal interference color hues due to variation in thickness, determined by the underlying microstructure. When this variation of electrolytic etching occurs, it is referred to as *anodizing.* Comparable nonelectrolytic processes are heat tinting and deposit etching.

Potentiostatic etching is an advanced form of electrolytic etching, which produces the ultimate etching contrast through highly controlled conditions. The potential of the specimen, which would usually change with changes of electrolyte concentration, is maintained at a fixed level through the use of a potentiostat and suitable reference standards. Clearly pronounced contrast can be obtained with this method where this is otherwise not possible. In some cases, the cell current can be maintained with a coulombmeter to determine the extent of etching (controlled etching).

On completion of any chemical or electrochemical etching process, the specimen should be rinsed in clean water to remove the chemicals and stop any reactions from proceeding further. Sometimes, for example, in etching for segregations using the Oberhoffer method, it is advisable to rinse in alcohol first. Otherwise, copper could precipitate on the specimen surface because of the change in the degree of dissociation. After specimens are water rinsed, they should be rinsed in alcohol and dried in a stream of warm air. The use of alcohol speeds up the drying action and prevents the formation of water spots. If etching produces water-soluble layers, water must be avoided in the rinsing step. Mounted specimens must be cleaned thoroughly to avoid the destructive effects of etchants and solvents seeping from pores, cracks, or clamp interfaces.

It may be helpful to use an ultrasonic cleaner to avoid these problems. If specimens are of a high porosity or if highly concentrated acids are used for etching (as for example, for deep etching), it is advisable to neutralize the chemicals before rinsing and drying the specimen.

Physical Etching

Basic physical phenomena are also often used to develop structural contrast, mainly when conventional chemical or electrolytic techniques fail. They have the advantage of leaving surfaces free from chemical residues and also offer advantages where electrochemical etching is difficult—for example, when there is an extremely large difference in electrochemical potential between microstructural elements, or when chemical etchants produce ruinous stains or residues. Some probable applications of these methods are plated layers, welds joining highly dissimilar materials, porous materials, and ceramics.

Cathodic vacuum etching, also referred to as ion etching, produces structural contrast by selective removal of atoms from the sample surface. This is accomplished by using high-energy ions (such as argon) accelerated by voltages of 1 to 10 kV. Individual atoms are removed at various rates, depending on the microstructural details such as crystal orientation of the individual grains, grain boundaries, etc.

Thermal etching, used in high-temperature (hot-stage) microscopy, is also partly based on atoms leaving the material surface, as a result of additional energy. The predominant force in thermal etching, however, is the formation of slightly curved equilibrium surfaces having individual grains with minimum surface tension.

Structural contrast produced by *evaporated interference layers* is often considered to be an optical method. Since no modification in the optical path of the microscope is made, it is considered here as a physical technique. This is rational since the polished specimen is treated in a way which distinguishes it from optical etching. The surface of the specimen is coated under vacuum with an evaporated layer of material, producing interference effects. High refractive index materials, such as ZnSe, TiO_2, etc., are commonly used. The effect of the evaporated interference layer is caused by multiple light reflection produced at the interface between object and evaporated layer (Fig. 19).

Gas-reaction chambers are a recent development which permit precise control of evaporation during simultaneous microscopic observation.

Specimen Storage

When polished and etched specimens are to be stored for long periods of time, they must be protected from atmospheric corrosion. Desiccators and desiccator cabinets are the most common means of specimen storage, although plastic coatings and cellophane tape are sometimes used.

Reproducibility in Etching

For the most part, metallographic etching continues to be an empirical method with overtones of black magic. This condition is the result of the abundance of etching methods, nonuniform nomenclature and, frequently, the lack of knowledge concerning etchant mechanisms. For these reasons, it is difficult to present a clear review of etching processes.

Conventional etching, in particular, is difficult to reproduce, regardless of its simplicity. During the electrochemical processes, numerous side effects must be taken into account. For example, changes in the electrolyte and inhibiting reactions at the specimen surface must be considered, which cause polarization phenomena, overpotential, etc.

With the goal of achieving more reproducible and dependable structural contrast, various new methods have been developed in recent years. *Electrolytic potentiostatic etching, ion etching,* and *evaporation of interference layers* are advanced methods gradually becoming accepted. The principles by which these techniques operate are illustrated in Fig. 19. In the first two methods, the microstructure becomes visible through selective removal of the surface. In the last method, the contrast is produced by multiple reflection of the light beam in the evaporated layer of a material with high refractive index.

The development of more reproducible and contrasting etching methods is of particular importance for *quantitative image analysis.* These instruments are used to automatically determine the area fraction of various phases and are not sensitive to subtle differences. Therefore, sharply reproducible etching contrast is necessary to obtain accurate information.

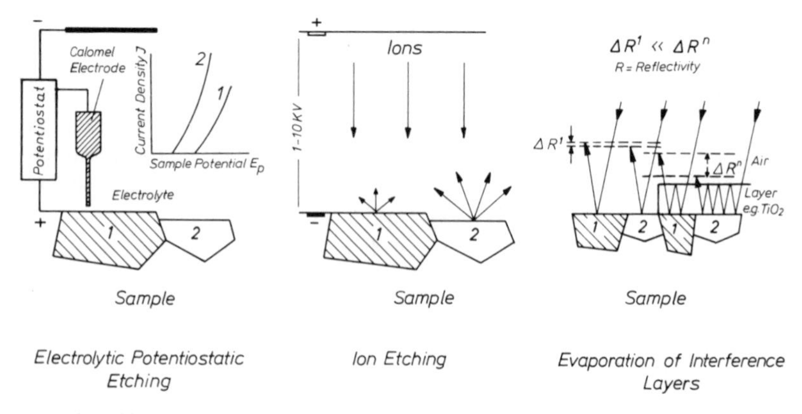

Fig. 19. Three methods for reproducible metallographic etching.

Etching Nomenclature

A classification of the most commonly used terms is possible on the basis of distinctive features. A distinction has already been made between optical, electrochemical, and physical etching on the basis of the kinetic phenomena occurring at the sample surface. Further distinctions are listed in Table 3, and explained in the next section. Often, terms are used which refer to the major component of the etchant (for example, dilute nitric acid, aqua regia, sodium thiosulfate, etc.), to the originator of the etchant (for example, Oberhoffer, Vilella, Murakami, Heyn, etc.), or to alloys or chief constituents for which the etchant is intended (for example, carbide, phosphide, steel etchant, etc.).

The multiplicity of terms is confusing because several different terms may apply to the same phenomenon. The same phenomena can be named according to different viewpoints, without converse limitation or exclusion of notation. Steels, for example, may be etched using either chemical or electrochemical methods; the microstructure may be revealed either by dissolution or precipitation phenomena; it may display distinct grain boundaries and/or colored grains, microstructures and/or macrostructures; combinations of etching procedures may be used, usually in order of increasing severity. Steel etching, therefore, can be generated by practically any of the etching methods, using one of the many suitable etchants under one of the many suitable conditions. All have one ultimate goal: to produce a structure with sufficient contrast to delineate as much detail as possible. Table 3 summarizes the etching terminology and how the terms are related.

Explanation of Etching Terms

anodic etching. Development of microstructure by selective dissolution of the polished surface under application of a direct current. Variation with layer formation: anodizing.

attack-polishing. Simultaneous etching and mechanical polishing.

cathodic etching. See *ion etching.*

cold etching. Development of microstructure at room temperature and below.

controlled etching. Electrolytic etching with selection of suitable etchant and voltage resulting in a balance between current and dissolved metal ions.

crystal-figure etching. Discontinuity in etching depending on crystal orientation. Distinctive sectional figures form at polished surface. Closely related to *dislocation etching.*

deep etching. Macroetching, especially for steels, to determine the over-all character of the material (presence of imperfections such as seams, forging bursts, shrinkage-void remnants, cracks, coring).

dislocation etching. Etching of exit points of dislocations on a surface. Depends on the strain field ranging over a distance of several atoms. Crystal figures (etch pits) are formed at exit points. For example, etch pits for cubic materials are cube faces.

Table 3. Relationships and Possible Variations in Metallographic Etching

I	1. Dissolution etching	1
	2. Precipitation etching	−2
II	3. Wet etching	++3
	4. Dry etching	++−4
III	5. Short-term etching	++++5
	6. Long-term etching	++++−6
	7. Cold etching	++++++7
	8. Hot etching	++++++−8
IV	9. Microetching	++++++++9
	10. Macroetching	++++++++−10
V	11. Immersion etching	+++−++++++11
	12. Drop etching	+++−++++++−12
	13. Etch rinsing	+++−++++++−−13
	14. Immersion etching, cyclic	+++−+++++++−−14
	15. Swabbing	++++++++++−−−−15
	16. Attack-polishing	++++++++++−−−−+16
	17. Anodic etching	+++−++++++−−−−++17
	18. Cathodic (ion) etching	+−−+++++++−−−−−−−18
	19. Heat tinting	−+−+++−+++−−−−−−−−19
	20. Thermal etching	−−−+++−+++−−−−−−−−−20
	21. Potentiostatic etching	+++−+++++−−−−−−−+−−−21
	22. Printing	+−+−++++−+−−−−−−+−−−−22
	23. Staining	−+++++++++++++−−+−+−−−23
	24. Double etching	++++++++++++++++++++−++24
	25. Multiple etching	++++++++++++++++++++−+++25
	26. Identification etching	+++++++++++++++++++−+++++26
	27. Controlled etching	+−+−++++++−−−−−−+−−−+−−−−+27
	28. Deep etching	+−+−++++−++++++++−−−−−−−−−+28
VI	29. Grain boundary etching	+++++++++++++++++++++−+++++−29
	30. Grain contrast etching	+++++++++++++++++++−+−+++++++30
	31. Crystal figure etching	+−+++++++−++++++++−+−−−++++−−+31
	32. Dislocation etching	+−+++++++−+++++++−−+−−−++++−+++32
	33. Network etching	+++++++++−++++++++++−−+++++−++++33
	34. Shrink etching	−++−+++++−+++−−−+−+−+−−−−−−−++−−−34
	35. Eutectic cell etching	++++++++++++++++++−+−++++++++−−−−−35
	36. Primary etching	++++++++−++++++−++++−+++−−++++−−−++36
	37. Secondary etching	+++++++++−+++++−++++−−+++++−+++++++−37

I	Changes in microsection surface
II	State of aggregation of etchant
III	Etching conditions (time and temperature)
IV	Used magnifications (macro, >25×; micro, <25×)
V	Etching methods and techniques
VI	Etching phenomena dependent upon microstructure
+	Combination of terms possible and existing
−	Terms are mutually exclusive or their combination is irrational

dissolution etching. Development of microstructure by surface removal.

double etching. Use of two etching solutions in sequence. The second etchant stresses a particular microstructural feature.

drop etching. Placing of a drop of etchant on the polished surface. Suitable for precious etchants.

dry etching. Development of microstructure under the influence of gases.

electrochemical (chemical) etching. General expression for all developments of microstructure through reduction and oxidation processes (redox reactions).

electrolytic etching. See *anodic etching.*

etch rinsing. Pouring etchant over tilted surface until desired degree of attack is achieved. Used for etchants with severe gas formation.

eutectic-cell etching. Development of eutectic cells (grains).

grain-boundary etching. Development of intersections of grain faces with the polished surface. Because of severe, localized crystal deformation, grain boundaries have higher dissolution potential than grains themselves. Accumulation of impurities in grain boundaries increases this effect.

grain-contrast etching. Development of grain surfaces lying in the polished surface of the microsection. These become visible through differences in reflectivity caused by reaction products on the surface or by differences in roughness.

heat tinting. Formation of colors (interference colors) in air or other gases, mostly at elevated temperature.

hot etching. Development and stabilization of the microstructure at elevated temperature in etchants or gases.

identification etching. Etching to expose particular microconstituents; all others remain unaffected.

immersion etching. Method in which a microsection is dipped into etching solution face up and is moved around during etching. This is the most common etching method.

immersion etching, cyclic. Alternate immersion into two etchants: (1) actual etchant; (2) solution used to dissolve layer formed during process 1.

ion etching. Surface removal by bombarding with accelerated ions in vacuum (1 to 10 kV).

long-term etching. Etching times of a few minutes to hours.

macroetching. Development of macrostructure for viewing with the unaided eye, or under lens magnifications up to 25× (50× in Europe).

microetching. Development of microstructure for microscopic examination. Usual magnification of more than 25× (50× in Europe).

multiple etching. Treatment of microsection sequentially with specific reagents attacking distinct microconstituents.

network etching. Formation of networks, especially in mild steels, after etching in nitric acid. These networks relate to subgrain boundaries.

optical etching. Development of microstructure under application of special illumination techniques (dark field, phase contrast, interference contrast, polarized light).

physical etching. Development of microstructure through removal of atoms from surface or lowering the grain-surface potential.

potentiostatic etching. Anodic development of microstructure at a constant potential. By adjusting the potential, a defined etching of singular phases is possible.

precipitation etching. Development of microstructure through formation of reac-

tion products at the surface of the microsection. (See also *staining.*)
primary etching. Development of cast structures including coring.
printing. A method in which a carrier material is soaked with an etchant and pressed against the surface of the specimen. The etchant reacts with one of the phases. Substances form which react with the carrier material. These leave behind a life-size image. Used for exposing particular elements—for example, sulfur (sulfur prints).
secondary etching. Development of microstructures deviating from primary structure through transformation and heat treatment in the solid state.
segregation (coring) etching. Development of segregation (coring) mainly in macrostructures and microstructures of castings.
short-term etching. Etching times of seconds to a few minutes.
shrink etching. Precipitation on grain surfaces. Shrinkage takes place during drying, which causes a cracking of the layer formed during etching. Crack orientation depends on the underlying structure.
staining. *Precipitation etching* that causes contrast by distinctive staining of microconstituents; different interference colors originate from surface layers of varying thickness. Proof of inhomogeneities.
strain etching. Etching that renders information on deformed and undeformed areas if present side by side. In strained areas, more compounds are precipitated.
swabbing. Wiping of the specimen surface with a cotton ball saturated with etchant to simultaneously remove reaction products.
thermal etching. Annealing of the specimen in vacuum or inert atmosphere. Used primarily in high-temperature microscopy.
tinting. See *heat tinting.*
wet etching. Development of microstructure with liquids (acids, bases, neutral solutions, mixtures of solutions).
wipe etching. See *swabbing.*

Nondestructive Metallographic Testing

Large structures or machines may be tested metallographically without destructive preparation. This may be accomplished by using portable devices which are either electrolytic or abrasive in principle. Surfaces prepared in this manner must then be examined at the site by using a portable microscope placed on, or attached to, the object.

Replicas are sometimes an advantage because they may be made *in situ,* and then taken to the laboratory for observation under more favorable conditions. With this technique, pieces of cellulose acetate or nitrate are softened on one face with a suitable solvent and pressed against a polished surface. The soft plastic surface flows to conform to the features of the sample. When it is dry, the plastic replica is removed from the surface and taken to the laboratory for viewing by transmitted and incident light. For increased contrast, carbon or metal may be evaporated onto the surface to produce shadowing. Another method of producing replicas employs an aluminum strip coated with plastic.

This product yields excellent image quality with no preparation other than the initial contact step.

Other applications of the replication technique include determinations of surface roughness and the making of observations in otherwise inaccessible areas. Radioactive materials are a particularly interesting application, since the replicas are practically unaffected by the radiation and can be handled with greater safety. As soon as the replicas are moved from the otherwise hazardous environment, they are much safer to use. Fractures are still another useful application, although considerable care must be taken when stripping replicas from coarse fracture surfaces. Curved surfaces can be handled more easily because the replica may be flattened out for use under the microscope.

Chapter 2:
Preparation of Metals and Alloys

The preparation procedures and etching reagents described in the tables that follow have been used successfully in many applications. Naturally, some modifications may be desirable or necessary under certain circumstances. Because of normal variations in any individual technique and the many variables in the processes, even proven etchants may fail. The metallographer must be prepared to make any adjustments required to find a satisfactory solution to his immediate problem. Etchant compositions which contain directions within parentheses—for example, "(concentration is variable)" —may require some experimentation to determine the exact formula. The general material headings are only a guideline for locating the best family of etchants. Wide variations in alloy composition within these areas will strongly affect the final choice of etching reagents. Only etchants with proven instructions for use have been listed.

Many of the listed reagents and test materials are hazardous; these are identified by three exclamation marks (!!!). Appendix A contains more extensive directions, which must be carefully observed.

Silver

Ag

Silver is extremely ductile. Deeply damaged layers are easily formed during preparation of microsections. Abrasive particles and polishing compounds tend to become embedded in the surface. Care must be taken during mechanical grinding and polishing. Silver forms water-insoluble halides, hydroxides, and carbonates (avoid tap water). Silver alloys usually do not cause as much difficulty. Silver compounds are often explosive, particularly in the dry state.

Macroetching

Preparation:

Wet grinding (using distilled water) on already used silicon carbide papers down to 600 grit (coating the paper with wax possibly helps).

Etching:

No.	Etchant		Conditions	Remarks
Ag M1	Methanol (95%) Nitric acid (1.40)	95 ml 10 ml	Several min.	Pure Ag and low-alloy Ag. Grain contrast.

Microetching

Preparation:

Grinding:

a. Wet on silicon carbide papers down to 600 grit (coat papers with wax; use distilled water).

b. Prepare by microtome, using cutters of cemented carbides or, better still, diamond. Also useful for all silver alloys with a hardness below 150 HV.

Polishing:

a. Use alumina or diamond paste down to finest particle size. An etch-polishing sequence or multiple polishing, using one of the listed microetchants, helps to decrease the deformed layer.

b. Electrolytic:

1. Dist. water	1000 ml	10 min,	!!! See Appendix A.
Potassium cyanide	37 g	2.5–3 V dc,	
Potassium carbonate	38 g	Ag cathode.	
Silver cyanide	35 g		
2. Dist. water	200 ml	4–6 min, approx 40 °C	
Ethanol (96%)	380 ml	(100 °F),	
Phosphoric acid (1.71)	400 ml	2.5–3 V dc,	
		Al cathode.	
3. Dist. water	1000 ml	Approx 1 min,	!!! See Appendix A.
Sodium cyanide	100 g	2.5 V dc,	
Potassium ferrocyanide	100 g	graphite cathode.	

c. Electrolytic lapping:

Dist. water	1000 ml	2–5 min,
Sodium thiosulfate	12 g	6–8 mA/cm^2.
Thiourea	2 g	
Copper (II) nitrate	1 g	

d. Chemical:

Dist. water	800 ml	!!! See Appendix A.
Sat. aqueous sol. of chromium (VI) oxide	100 ml	
Hydrochloric acid (10%)	45 ml	

Etching:

No.	Etchant		Conditions	Remarks
Ag m1	Dist. water Sulfuric acid (1.84) Chromium (VI) oxide (Concentration variable)	100 ml 2–11 ml 2 g	Up to 1 min. !!! See Appendix A.	Predominantly used for Ag alloys, especially Ag-Ni alloys and Ag-Mg-Ni alloys.
Ag m2	Sulfuric acid (1.84) Sat. aqueous sol. of potassium dichromate Sat. aqueous sol. of sodium chloride	10 ml 100 ml 2 ml	Secs to mins, 1:9 dilution with dist. water. Possibly without sulfuric acid.	Pure Ag and Ag alloys. Ag solders.

Ag m3	a. Dist. water Ammonium persulfate b. Dist. water Potassium cyanide	100 ml 10 g 100 ml 10 g	30 s to 2 min. Mix (a) and (b) immediately before etching in proportion 1:1. !!! See Appendix A.	Ag and low-alloy Ag.
Ag m4	Ammonia water Potassium cyanide	100 ml 5–10 g	Several secs. !!! See Appendix A.	Pure Ag and Ag composites with other metals.
Ag m5	Ammonia water Hydrogen peroxide (3%)	50 ml 50 ml	Up to 1 min. Use fresh!	Pure Ag. Ag-Ni alloys. Ag-Pd alloys.
Ag m6	Dist. water Hydrogen peroxide (3%) Ammonia water	25 ml 50 ml 25 ml	Up to 1 min. Use fresh!	Ag-rich Ag-Cd alloys. Ag solders. Ag-Cu alloys.
Ag m7	Dist. water Iron (III) chloride	100 ml 2 g	5–30 s.	Ag solders.
Ag m8	Aqueous sol. of sodium hydroxide (10%) Aqueous sol. of potassium ferricyanide (30%)	10 ml 10 ml	5–15 s. If attack too rapid, dilute 50% with dist. water.	Ag-Mo alloys. Ag-W alloys. Ag-W carbides.
Ag m9	*Electrolytic:* Dist. water Citric acid	10 ml 10 g	15 s to 1 min. 6 V dc, Ag cathode. Possibly 2–3 drops nitric acid (1.40).	Ag alloys.

Aluminum

Al

Aluminum and many of its alloys are soft and ductile and have a tendency to form deformed layers. Grinding and polishing compounds are easily embedded. Anodic layer, lacquers, galvanic layers, and soldered, brazed, or welded joints must be examined unetched.

Macroetching

Preparation:

Coarse grinding is normally sufficient (wet on silicon carbide paper).

Etching:

No.	Etchant		Conditions	Remarks
Al M1	Dist. water Sodium hydroxide	100 ml 10–20 g	5–15 min, 60–70 °C (140–160 °F). More concentrated solutions work also at room temperature.	Suitable for most types of Al and Al alloys.

Al M2	Hydrochloric acid (1.19) Nitric acid (1.40) Hydrofluoric acid (40%)	75 ml 25 ml 5 ml	Secs to mins. Use fresh! Possibly diluted with 25 ml dist. water. Rinse in warm water. !!! See Appendix A.	Al alloys containing Cu, Mn, Si, Mg, Ti. Al cast alloys with high Si content.
Al M3	Dist. water Hydrochloric acid (1.19) Hydrofluoric acid (40%) (Flick's reagent)	90 ml 15 ml 10 ml	5 s to 3 min. !!! See Appendix A.	Al-base materials in general. Pure Al. Cu-Al alloys.
Al M4	Dist. water Hydrochloric acid (1.19) Nitric acid (1.40) Hydrofluoric acid (40%) (Concentration variable) (Keller's reagent)	20 (50) ml 20 (15) ml 20 (25) ml 5 (10) ml	1–3 min. !!! See Appendix A.	Al-base materials in general. High-purity Al. Alloys of Al-Mn, Al-Mg, Al-Mg-Mn, and Al-Mg-Si. Grain size. Rolling direction. Welded joints. Also useful for microetching.
Al M5	Dist. water Hydrochloric acid (1.19) Nitric acid (1.40) Hydrofluoric acid (40%) (Tucker's reagent)	25 ml 45 ml 15 ml 15 ml	Secs to mins. Use fresh! !!! See Appendix A.	High-purity Al. Mn-Al, Si-Al, Mg-Al, and Mg-Si-Al alloys.
Al M6	Dist. water Sulfuric acid (1.84)	100 ml 5.5 ml	Secs to mins.	Surface imperfections of pure Al.

If macroetching results in a coated surface, the coating can be removed by a strong solution of nitric acid.

Microetching

Preparation:

Grinding:

Wet on silicon carbide papers down to 600 grit.

Polishing:

a. With diamond paste (particle size 6-1 μm, possibly 0.25 μm), followed by 1–5 min on a soft cloth with a suspension: 120 ml hot distilled water, 20 ml aqueous solution of ammonium tartrate (5%), and 1 g magnesium oxide. Suspension is filtered prior to use. Dense nylon or cotton webbing is a suitable filter material. Copper-free samples should not be polished on the same cloth as samples containing copper (formation of dark precipitate may occur).

b. With diamond paste (particle size 6 μm).
   Then electrolytically:

   1. Dist. water 250 ml; Ethanol (96%) 380 ml; Phosphoric acid (1.71) 400 ml — 4–6 min, 40 °C (100 °F), 40–60 V dc, Al cathode.

1. Dist. water Ethanol (96%) Phosphoric acid (1.71)	250 ml 380 ml 400 ml	4–6 min, 40 °C (100 °F), 40–60 V dc, Al cathode.

2. Methanol (95%) Nitric acid (1.40) Perchloric acid (60%)	950 ml 15 ml 50 ml	10–60 s, 30–60 V dc, stainless steel cathode. !!! See Appendix A.
3. Dist. water Ethanol (96%) Perchloric acid (60%)	140 ml 800 ml 60 ml	15–60 s, 30–80 V dc, stainless steel cathode. !!! See Appendix A.
4. Dist. water Glycerine Sulfuric acid (1.84)	220 ml 200 ml 580 ml	1–2 min, 1.5–12 V dc, stainless steel cathode.

c. With diamond paste (particle size 6 μm).
Then chemically:

1. Sulfuric acid (1.84) Phosphoric acid (1.71) Nitric acid (1.40)	25 ml 70 ml 5 ml	30 s to 2 min, 85 °C (185 °F).
2. Dist. water Sat. aqueous sol. of ammonium hydrogen fluoride Nitric acid (1.40)	65–80 ml 10–20 ml 10–20 ml	5–30 s, 50–60 °C (120–140 °F).

d. With alumina slurry No. 1 (5 μm) and No. 1C (1 μm; Linde C) on a fast-running felt-covered wheel, followed by alumina slurry No. 2 (0.3 μm; Linde A) on cotton velvet. For final polishing, alumina slurry No. 3 (0.05 μm gamma alumina; Linde B) diluted with alcohol on silk velvet.

e. Electrolytic lapping:

Dist. water No. 3 alumina slurry (0.05 μm) Sulfuric acid (1.84) Sodium fluoride	95 ml 10 ml 5 ml 0.5 g	2–6 min, 1–10 mA/6.5 cm^2, dc. Change polarity every 1.5 s. Velvet cloth.	Pure Al. Duralumin. Al with small additions of Zn, Mg, Cu, Fe, Ti, Mn, Si. Al-Si alloys up to 18%.

f. Milling with diamond cutting tool. Possibly followed by (b). Methanol for lubrication and cooling; only for harder alloys.

Etching:

No.	Reagent		Conditions	Remarks
Al m1	Dist. water Hydrofluoric acid (40%) If necessary, increase content of hydrofluoric acid to 10 ml.	100 ml 0.5 ml	10–60 s. !!! See Appendix A. Minutes to 1 h. Alternate polishing and etching.	For most types of Al and Al alloys. Grain boundaries. Slip lines in high-purity Al.
Al m2	Dist. water Nitric acid (1.40) Hydrochloric acid (1.19) Hydrofluoric acid (40%) (Concentration variable) (Dix-Keller's reagent)	190 (190) ml 5 (10) ml 3 (6) ml 2 (4) ml	10–30 s. Use fresh only! !!! See Appendix A.	For most types of Al and Al alloys. Exception: Al alloys with high Si content.

Al m3	Dist. water Sodium hydroxide	100 ml 1–2 g	a. 5–10 s, possibly at 50 °C (120 °F). Use fresh only! Rinse in mixture: 95 ml dist. water, 5 ml nitric acid (1.40). b. Etch 15 min, rinse in water approx. 10 min.	Pure Al. Alloys of Cu-Al, Mg-Si-Al, Mg-Al, and Zn-Al. If etching times are short, Al_2Cu is not colored. For shrink etching.
Al m4	Dist. water Sodium hydroxide Zinc chloride	100 ml 0.5–25 g 1 g	Few secs to mins.	For most types of Al and Al alloys.
Al m5	Dist. water Nitric acid (1.40)	75 ml 25 ml	40 s, 70 °C (160 °F).	For most types of Al and Al alloys, especially those containing Cu.
Al m6	Dist. water Nitric acid (1.40) Hydrofluoric acid (40%) (Kroll's reagent)	92 ml 6 ml 2 ml	15 s. !!! See Appendix A.	Especially for Cu-Al alloys. (Also useful in macroetching).
Al m7	Dist. water Sulfuric acid (1.84)	80 ml 20 ml	30 s to 3 min, 70 °C (160 °F).	Differentiation of intermediate phases in Al alloys with high contents of Cu, Mn, Mg, Fe, Be, Ti. Fe_3Al.
Al m8	Dist. water Phosphoric acid (1.88 crystal)	100 ml 9 g	30 min.	Detection of susceptibility to intercrystalline corrosion of Al-Mg-base alloys. Distinct grain boundaries in corrosion-sensitive alloys.
Al m9	Same as Al m8		Anneal sample to 100 °C (212 °F) and slow cool before etching.	Precipitates of beta-Al_8Mg_5 or Al_3Mg_2, respectively. Shows strain in Al-Mg-base alloys.
Al m10	Methanol (95%) Hydrochloric acid (1.19) Nitric acid (1.40) Hydrofluoric acid (40%)	25 (50) ml 25 (30) ml 25 (20) ml 1 drop	10–60 s. !!! See Appendix A.	Pure Al. Al-Mg and Al-Mg-Si alloys. Etch figures.
Al m11	Dist. water Sodium hydroxide Potassium ferricyanide	60 ml 10 g 5 g	2 min	Precipitates. Grain boundaries. Al-Si and Al-Cu alloys.

Al m12	Dist. water Phosphoric acid (1.71)	90 ml 10 ml	a. 1–3 min, 50 °C (120 °F). b. *Electrolytic:* 5–10 s, 1–8 V dc, stainless steel cathode	Pure Al. Al-Cu, Al-Mn, Al-Mg, Al-Mg-Si alloys.
Al m13	*Electrolytic:* Dist. water Fluoboric acid (35%) (Barker's reagent)	 200 ml 10 ml	1–2 min, 20–40 V dc. Do not wipe during rinsing. Cathode and anode fixtures should be made of high-purity Al, Pb, or stainless steel.	Pure Al. Al-Zn, Al-Mn, Al-Mg-Si, Al-Zn-Mg, Al-Mn-Mg alloys. Difficult with Al-Cu-Mg alloys. View in polarized light.

Gold, Iridium, Osmium, Palladium, Platinum, Rhodium, Ruthenium

Au | Ir | Os | Pd | Pt | Rh | Ru

These metals are usually soft. They have a tendency to produce deformed layers (although iridium, osmium, rhodium and ruthenium represent exceptions depending on purity). The alloys do not pose particular difficulties in preparing metallographic specimens.

Macroetching

Preparation:

Specimen preparation for macroetching is identical to that detailed below for microetching.

Etching:

No.	Etchant		Conditions	Remarks
Au M1	Hydrochloric acid (1.19) Nitric acid (1.40)	66 ml 34 ml	Few mins. Use hot! Use fresh! !!! See Appendix A.	Au. Pt alloys. Pd alloys.
Au M2	Lactic acid (90%) or Hydrochloric acid (1.19) Nitric acid (1.40) Hydrofluoric acid (40%)	50 ml 20 ml 30 ml	Few mins. !!! See Appendix A.	Ru and Ru alloys. Os and Os alloys. Rh and Rh alloys.
Au M3	*Electrolytic:* Sat. aqueous sol. of sodium chloride Hydrochloric acid (1.19)	 80 ml 20 ml	Few mins, 6 V dc, Pt cathode.	Pt and Pt alloys.

Microetching

Preparation:

Grinding:

a. Wet on silicon carbide papers down to 600 grit (coat papers with wax).
b. Prepare by microtome using diamond knife. Also useful for all alloys with hardness below 150 HV.

Polishing:

a. Alumina slurry or diamond paste down to finest particle size. Multiple polishing with one of the microetchants listed for electrolytic etching eliminates the deformed layer more quickly.
b. Electrolytic:

1. Dist. water Potassium cyanide Potassium carbonate Gold (III) chloride	1000 ml 80 g 40 g 50 g	2–4 min, 7.5 V dc, graphite cathode. !!! See Appendix A.	Mostly Au.
2. Dist. Water Glacial acetic acid Chromium (VI) oxide	3 ml 75 ml 15 g	8–12 V dc, stainless steel cathode. Mixing of electrolyte at 65 °C (150 °F), 1 h. !!! See Appendix A.	Mostly Au.

c. Electrolytic lapping:

1. Dist. water Sodium thiosulfate Potassium thiocyanate Ammonium chloride	820 ml 50 g 70 g 60 g	2–3 min, 1.2–1.5 A/cm² dc, 150 rpm. !!! See Appendix A.	Pd and Pd alloys. Au-Pt 90 alloy.
2. Dist. water Potassium thiocyanate Ammonium chloride Ammonium thiosulfate	940 ml 20 g 20 g 20 g	4–5 min, 55 mA/cm² dc. !!! See Appendix A.	Au.
3. Dist. water Hydrochloric acid (1.19) Sodium fluoride Potassium chloride Potassium nitrate Potassium thiocyanate	610 ml 50 ml 150 g 100 g 50 g 40 g	Approx 15 min, 0.5 A/0.25 cm² ac. !!! See Appendix A.	Pt. Au-Pt alloys 70/30 and 50/50.

Etching:

No.	Etchant		Conditions	Remarks
Au ml	Nitric acid (1.40) Hydrochloric acid (1.19) (Concentration variable)	40 (1) ml 60 (10) ml	Few secs to 1 min. May require heating. Use fresh only! !!! See Appendix A.	Pure Au and Pd. Au-Pt, Pd alloys with more than 90% content of precious metals. Rh alloys.

Au m2	a. Dist. water Potassium cyanide b. Dist. water Ammonium persulfate	100 ml 10 g 100 ml 10 g	30 s to 2 min. (a) and (b) are mixed in proportion 1:1 before use. Double amount of potassium cyanide and ammonium persulfate. !!! See Appendix A.	Pd, Pt. Au alloys with less than 90% content of precious metals. For Au alloys with high content of precious metals. White gold. Pd and Pt alloys.
Au m3	Dist. water Hydrogen peroxide (3%) Iron (III) chloride	100 ml 100 ml 32 g	Secs to mins.	Au-Cu-Ag alloys.
Au m4	Dist. water Hydrochloric acid (1.19) Nitric acid (1.40) (Concentration variable)	30 (50) ml 25 (100) ml 5 (10) ml	1–5 min. Use hot! Remove precipitate of gold chloride with ammonia water.	Pure Pt and Pd. Au alloys. Proportions in parentheses especially useful for Pt.
Au m5	Hydrochloric acid (1.19) Chromium (VI) oxide	100 ml 1–5 g	Secs to mins. !!! See Appendix A.	Pure Au and Au-rich alloys. Pd and Pd alloys.
Au m6	Dist. water Potassium ferricyanide Sodium hydroxide	150 ml 3.5 g 1 g	Few mins. !!! See Appendix A.	Os and Os-W alloys.
Au m7	Dist. water Hydrochloric acid (1.19) Hydrogen peroxide (3%)	80 ml 20 ml 1 ml	Few mins.	Ru-rich alloys. Ru-Mo alloys.
Au m8	*Electrolytic:* Dist. water Potassium cyanide	 100 ml 5 g	 1–2 min, 1–5 V ac, 0.5–1.5 A/cm^2, Pt cathode. !!! See Appendix A.	 Pt and Pt alloys. Au and Au alloys.
Au m9	*Electrolytic:* Dist. water Hydrochloric acid (1.19) Sodium chloride	 65 ml 20 ml 25 g	 25 sec, 10 V ac. 1 min, 1.5 V ac. 1–2 min, 20 V ac. 1 min, 6 V ac. 1 min, 5–20 V ac. Graphite or Pt cathode.	 Rh-base alloys. Pt-10% Rh alloys. Ir alloys. Pure Pt and Pt alloys. Ru-base alloys.
Au m10	*Electrolytic:* Ethanol (96%) Hydrochloric acid (1.19)	 90 ml 10 ml	 30 s, 10 V dc, graphite cathode.	 Os-base alloys. Pure Pd and Pd alloys. Pt-Au alloys. Ir.
Au m11	*Electrolytic:* Hydrochloric acid (1.19)		 1–2 min, 5 V ac, Pt cathode.	 Rh-base alloys. Au and Pt. Grain-contrast etch.

	Etchant		Conditions	Remarks
Au m12	*Electrolytic:* Dist. water Hydrochloric acid (1.19) (Concentration variable)	90 ml 5–10 ml	30 min to 3 h, 0.1 A/cm², graphite or Pt cathode.	Ir.
Au m13	*Electrolytic:* Dist. water Sulfuric acid (1.84)	80 ml 20 ml	Up to 1 h, 1–5 V ac, 0.05–0.2 A/cm², graphite cathode.	Pt alloys. Rh, Ir.

Be

Beryllium

Preparation steps generate toxic dust, which must be carried out in a glove box. Beryllium must never be inhaled. (!!! See Appendix A and references on safety and toxicology in Appendix C.)

Macroetching

Preparation:

Grinding:

Always wet on silicon carbide papers down to 600 grit, which is sufficient for coarse-grain material. Beryllium with small grain size must be polished.

Polishing:

Diamond paste (6–0.25 μm), followed by alumina suspension on a finely woven, sturdy cloth. Also valid for microetching.

Etching:

No.	Etchant		Conditions	Remarks
Be M1	Dist. water Hydrochloric acid (1.19) Ammonium chloride	90 ml 10 ml 4 g	Few mins. Immerse or swab.	Technical types of Be. Especially with large grain size.
Be M2	Dist. water Hydrochloric acid (1.19) Ammonium chloride Picric acid	90 ml 10 ml 2 g 2 g	Few mins. Immerse or swab.	Technical types of Be. Especially with large grain size. Low-alloy Be.
Be M3	Sulfuric acid (1.84) Phosphoric acid (1.71) Chromium (VI) oxide (Possibly some dist. water)	25 ml 500 ml 59 g	Secs to mins, 45 °C (110 °F) !!! See Appendix A.	Grain-boundary etch.

With fine-grain Be, macroetching is difficult.

Microetching

Preparation:

Compare technical tips for macroetching.

Polishing:

a. Concentrated slurry of magnesium oxide in hydrogen peroxide (30%) on billiard cloth.

b. Alumina slurry in oxalic acid (10%) on silk. Final polish (15–30 s) on nylon with long nap using No. 3 (0.05 μm, Linde B) alumina slurry in a solution of 100 ml water, 14 ml phosphoric acid (1.71), 1 ml sulfuric acid (1.84), and 20 g chromium (VI) oxide, diluted with 10 parts distilled water (!!! See Appendix A.). Solution is added in drops during polishing.

c. Electrolytic:

1. Dist. water	200 ml	250 A/cm^2,
Phosphoric acid (1.71)	900 ml	70–80 °C (160–180 °F),
Sat. sol. of chromium (VI)		stainless steel cathode.
oxide in dist. water	240 ml	!!! See Appendix A.
2. Phosphoric acid (1.71)	100 ml	2–4 A/cm^2,
Sulfuric acid (1.84)	30 ml	stainless steel cathode.
Glycerol	30 ml	!!! See Appendix A.
Ethanol (96%)	30 ml	

d. Electrolytic lapping:

Ethylene glycol (1.11)	97 ml	1–1.5 A/cm^2.
Hydrochloric acid (1.19)	1 ml	Nylon cloth.
Nitric acid (1.40)	2 ml	

Etching:

Beryllium is suitable for observations in polarized light.

No.	Etchant		Conditions	Remarks
Be m1	Dist. water or ethanol (96%) Hydrofluoric acid (40%)	100 ml 2–10 ml	Secs to mins. !!! See Appendix A.	Be alloys.
Be m2	Dist. water Sulfuric acid (1.84)	100 ml 5 ml	1–15 s.	For most types of Be and Be alloys.
Be m3	Dist. water Ammonia water Hydrogen peroxide (30%)	50 ml 20 ml 3 ml	Secs to mins. Use fresh only!	Be-Ag and Be-Al-Ti alloys.
Be m4	Glycerol Hydrofluoric acid (40%) Nitric acid (1.40)	25 ml 5 ml 5 ml	Approx 15 s. HF content may be raised to 15 ml. !!! See Appendix A.	Be and Be alloys.
Be m5	Lactic acid (90%) Nitric acid (1.40) Hydrofluoric acid (40%)	50 ml 50 ml 50 ml	Secs to mins. !!! See Appendix A.	Be-U, Be-Nb, Be-Y, and Be-Zr alloys.
Be m6	Dist. water Oxalic acid	100 ml 10 g	2 min. 16 min. Boiling.	Precipitates. Grain boundaries.
Be m7	Sat. sol of copper (II) sulfate in dist. water		30 s.	Precipitates.

Be m8	*Electrolytic:* Ethanol (96%)	35 ml	30–45 s, below	Be and Be alloys.
	Perchloric acid (10%)	10 ml	35 °C (95 °F), 50 V	Rapid attack.
	Butyl glycol	10 ml	dc, stainless steel cathode. !!! See Appendix A.	
Be m9	*Electrolytic:* Ethylene glycol (1.11)	294 ml	6 min,	Be and Be alloys.
	Hydrochloric acid (1.19)	4 ml	30 °C (85 °F), 13–20 V dc, stainless steel	
	Nitric acid (1.40)	2 ml	cathode.	
Be m10	*Electrolytic:* Phosphoric acid (1.71)	100 ml	1 min,	Be. Grain-boundary etch.
	Glycerol	30 ml	cool (10 °C; 50 °F),	Also used to increase
	Ethanol (96%)	30 ml	25 V dc, stainless	contrast in polarized
	Sulfuric acid (1.84)	2.5 ml	steel or Mo cathode.	light.

Bi

Sb

Bismuth and Antimony

Macroetching

Preparation:

Grinding:

Wet on silicon carbide papers down to 320 grit; if necessary, down to 600 grit.

Polishing:

a. See composition b1 for polishing aluminum in preparation for microetching. Also with alumina on velvet (approx. 150 rpm).

b. Electrolytic:

1.	Glycerol	75 ml	1–5 min,	Special for Bi.
	Glacial acetic acid	12.5 ml	12 V dc.	!!! See Appendix A.
	Nitric acid (1.40)	12.5 ml	Do not store!	
2.	Sat. aqueous sol. of potassium iodide	98 ml	30 s, several times, 5–7 V dc, 0.2 A/cm^2.	
	Hydrochloric acid (1.19)	2 ml		

Etching:

No.	Etchant		Conditions	Remarks
Bi M1	a. Dist. water	160 ml	First with (a) at	Sb-Pb, Bi-Sn alloys.
	Nitric acid (1.40)	40 ml	40 °C (100 °F). Re-	
	Glacial acetic acid	30 ml	polish until surface has become bright.	
	b. Dist. water	400 ml	Repeat etching with	
	Glacial acetic acid	1 ml	(b) 1–2 h	

No.	Etchant		Conditions	Remarks
Bi M2	a. Dist. water Nitric acid (1.40) b. Dist. water Ammonium molybdate	220 ml 80 ml 300 ml 45 g	Secs to mins. Equal amounts of (a) and (b) are mixed before use.	Technically pure Sb. Crystal arrangement. Sb-Bi alloys.
Bi M3	Dist. water Citric acid Ammonium molybdate	100 ml 25 g 10 g	Secs to mins.	Crystal arrangement. Casting imperfections in Sb and Bi.

Microetching

Preparation:

Same as for macroetching.

Etching:

No.	Etchant		Conditions	Remarks
Bi m1	Dist. water Hydrochloric acid (1.19) Hydrogen peroxide (30%)	70 ml 30 ml 5 ml	Secs to mins.	For pure and technically pure types of Sb. Low-alloy Sb.
Bi m2	Glacial acetic acid Hydrogen peroxide (30%)	30 ml 10 ml	Secs to mins.	Sb and Sb alloys.
Bi m3	Dist. water Hydrochloric acid (1.19) Iron (III) chloride	100 ml 30 ml 2 g	Secs to mins.	Sb and Sb alloys. Pb-Sb, Bi-Sn, and Bi-Cd alloys.
Bi m4	Dist. water Hydrochloric acid (1.19) Aqueous sol. of sodium thiosulfate (16%) Aqueous sol. of chromium (VI) oxide (10%) (Czochralski's etchant)	30 ml 15 ml 50 ml 3 ml	Add chromium (VI) oxide shortly before use. Chromium (VI) oxide content can be raised. !!! See Appendix A.	Sb alloys. Grain contrast etch.
Bi m5	Dist. water Hydrochloric acid (1.19)	50 ml 50 ml	1–10 min.	Sb, Bi, and their alloys.
Bi m6	Ethanol (96%) Nitric acid (1.40)	95 ml 5 ml	Secs to mins.	Bi-Sn eutectic. Bi-Cd alloys.
Bi m7	Glycerol Nitric acid (1.40) Glacial acetic acid (Concentration variable)	100 (100) ml 25 (9) ml 25 (9) ml	Secs to mins.	Sb-Pb alloys.
Bi m8	Dist. water Silver nitrate	100 ml 5 g	Secs to mins.	Bi.

Cd | In | Tl

Cadmium, Indium, Thallium

Cadmium, indium, and thallium, as well as many of their alloys, are soft and have a low recrystallization temperature. Pressure during grinding and polishing, and heating through friction should be kept as low as possible; otherwise, deformed layers may form or fictitious microstructures may be observed. Cadmium and thallium are toxic (!!! see Appendix A).

Microetching

Preparation:

Grinding:

a. Use a special file for light and soft metals.
b. Wet on SiC papers (coat with wax) down to grit size 600.
c. Use lathe, depth of cut below 25 μm.
d. Prepare by microtome using cutters of cemented carbides or, better still, diamond; eliminates polishing.

Polishing:

a. With alumina slurry or diamond paste down to finest particle size on rotating wheel, 200–300 rpm. If surface appears milky, add 1 g ammonium acetate to 1000 ml distilled water, and rinse specimen.
b. Electrolytic:

1. Dist. water Phosphoric acid (1.71)	550 ml 450 ml	30 min, 2 V dc, Ni cathode.	Cd, Tl.
2. Methanol (95%) Nitric acid (1.40)	670 ml 330 ml	1–2 min, 40–50 V dc, stainless steel cathode.	In.

c. Chemical:

Dist. water Nitric acid (1.40)	25 ml 75 ml	5–10 s.	Cd.

Etching:

Cd is suitable for observations in polarized light.

No.	Etchant		Conditions	Remarks
Cd m1	Ethanol (96%) Nitric acid (1.40)	98 ml 2 ml	Secs to mins.	Cd and Cd alloys. Tl.
Cd m2	Dist. water Chromium (VI) oxide	100 ml 10 g	1–10 min. !!! See Appendix A.	Cd and solder alloys containing Cd. Tl.
Cd m3	Dist. water Hydrochloric acid (1.19) Iron (III) chloride	100 ml 25 ml 8 g	Secs to mins.	Cd-Sn and Cd-Zn eutectics.
Cd m4	Dist. water Hydrofluoric acid (40%) Hydrogen peroxide (30%)	40 ml 10 ml 10 ml	5–10 s. !!! See Appendix A.	Cd, In, Tl. In-Sb and In-As alloys.

No.	Etchant		Conditions	Remarks
Cd m5	Ethanol (96%) Hydrochloric acid (1.19) Picric acid	100 ml 5 ml 1 g	Secs to mins.	In and In-rich alloys.
Cd m6	*Electrolytic:* Dist. water Glycerol Phosphoric acid (1.71)	 100 ml 200 ml 200 ml	 5–10 min, 8–9 V dc, Cd cathode.	 Cd, Tl.

Cobalt

Co

Pure cobalt is tough and has a tendency to produce deformed layers. Cobalt alloys do not usually pose these difficulties. The distinctly heterogeneous alloys, such as Stellite, have a tendency to form relief structures, since the hardness of the individual phases differs greatly.

Macroetching

Preparation:

Wet grinding on silicon carbide papers down to 600 grit.

Etching:

No.	Etchant		Conditions	Remarks
Co M1	Dist. water Hydrochloric acid (1.19)	50 ml 50 ml	30–60 min, hot. Rinse in hot water	Co-Cr alloys, also Stellite.
Co M2	Dist. water Nitric acid (1.40) Hydrochloric acid (1.19) Iron (III) chloride	100 ml 10 ml 50 ml 10 g	Swab. Rinse in warm water.	Co-25Cr-10Ni-8W, Co-21Cr-20Ni, and Co-3Cr-3Mo-1Nb alloys. Other Stellites.
Co M3	Dist. water Hydrochloric acid (1.19) Nitric acid (1.40)	25 ml 50 ml 25 ml	Secs to mins.	Co-Ni-Fe-base alloys. High-temperature alloys.

Microetching

Preparation:

Grinding:

Same as for macroetching. For cemented carbides use silicon carbide slurries (grit size 220–800) in distilled water on cast iron wheel.

Polishing:

a. Diamond paste (particle size 15-1 μm).

b. Electrolytic:

1. Methanol (95%) Nitric acid (1.40)	600 ml 330 ml	10–60 s, 40–70 V dc, stainless steel cathode.	!!! See Appendix A.
2. Phosphoric acid (1.71)		3–5 min, 1.5 V dc, stainless steel cathode.	
3. Dist. water Phosphoric acid (1.71)	600 ml 400 ml	1–15 min, 1–2 V dc, stainless steel or Co cathode.	

c. Chemical:

Lactic acid (90%)	40 ml	Secs to mins.
Hydrochloric acid (1.19)	30 ml	
Nitric acid (1.40)	5 ml	

Etching:

No.	Etchant		Conditions	Remarks
Co m1	Methanol (95%) Nitric acid (1.40)	100 ml 1–100 ml	Secs to mins. 5 min. !!! See Appendix A.	Pure Co. Co-Fe alloys. WC-TiC-NbC-Co – type cemented carbides.
Co m2	Dist. water Glacial acetic acid Hydrochloric acid (1.19) Nitric acid (1.40)	15 ml 15 ml 60 ml 15 ml	5–30 s. Wait at least 1 h before use. !!! See Appendix A.	Pure and low-alloy Co. Co-B alloys. Co-Ti alloys, and Co-Mn alloys. WC-TiC-TaC-Co – type cemented carbides. Grain-boundary etch.
Co m3	Hydrochloric acid (1.19) Hydrogen peroxide (30%)	100 ml 5 ml	Few secs. Use fresh only! !!! See Appendix A.	Wear-resistant alloys. Alloys used for Co-base cutting tools. Superalloys.
Co m4	Dist. water Ammonium persulfate	100 ml 20–30 g	Secs to mins.	Cemented carbides and other Co-base alloys.
Co m5	Dist. water Potassium hydroxide Potassium ferricyanide	100 ml 10 g 10 g	Secs to mins. !!! See Appendix A.	WC-TiC-TaC-Co and WC-NbC-Co – type cemented carbides.
Co m6	Hydrochloric acid (1.19) Nitric acid (1.40)	75 ml 25 ml	Secs to 5 min. Use fresh only. !!! See Appendix A.	WC-TiC-NbC-Co – type cemented carbides. Co-Pt alloys.
Co m7	Dist. water Hydrochloric acid (1.19) Methanol (95%) Nitric acid (1.40) Iron (III) chloride Copper (II) chloride	100 ml 100 ml 200 ml 5 ml 7 g 2 g	10–15 s. Immerse or swab.	Co-Fe alloys. Magnetic alloys.

Co m8	Dist. water Nitric acid (1.40)	100 ml 50 ml	90 s.	Co-Ga alloys.
Co m9	Hot etching in air		30 min, 400–450 °C (750–850 °F).	WC-TiC-(Ta, Nb)C-Co-type cemented carbides. Carbide phases are easily distinguishable from binder phases by differences in colors.
Co m10	Dist. water Glacial acetic acid Nitric acid (1.40)	100 ml 1 ml 1 ml	Few secs.	Co-Sm alloys.
Co m11	*Electrolytic:* Dist. water Potassium hydroxide Sodium carbonate	 100 ml 3 g 2 g	10 s, 3 V dc, stainless steel cathode. Increase of sodium carbonate content increases removal rate of WC phase. Deposit removed with diluted hydrochloric acid.	Cemented carbides with Co binder phase.
Co m12	*Electrolytic:* Dist. water Hydrochloric acid (1.19) Iron (III) chloride	 100 ml 5 ml 10 g	Few secs, 6 V dc, stainless steel cathode.	Pure Co. Co-Al alloys.
Co m13	*Electrolytic:* Dist. water or Hydrochloric acid (1.19) Chromium (VI) oxide	 100 ml 2–10 g	2–20 s, 3 V dc, stainless steel cathode. !!! See Appendix A.	Stellite up to 70% Co. Co-base superalloys. Co silicides.
Co m14	*Electrolytic:* Dist. water Hydrochloric acid (1.19)	 100 ml 5–10 ml	2–10 s, 3 V dc, graphite cathode.	Pure Co. Co-base superalloys.
Co m15	*Electrolytic:* Hydrochloric acid (1.19) Hydrogen peroxide (30%)	 100 ml 5 ml	3–5 s, 4 V dc, stainless steel cathode. !!! See Appendix A.	Co-base wear-resistant alloys and materials used for cutting tools. Superalloys.

Cr
Mo
Nb
Re
Ta
V
W

Chromium, Molybdenum, Niobium, Rhenium, Tantalum, Vanadium, Tungsten

In the pure state, these materials are soft and tough. In the commercially pure state, however, they are hard to brittle because of nonmetallic impurities.

Macroetching

Preparation:

Wet grinding on silicon carbide papers down to 600 grit.

Etching:

No.	Etchant		Conditions	Remarks
Cr M1	Dist. water Sulfuric acid (1.84)	90 ml 10 ml	2–5 min. Boiling.	Cr.
Cr M2	Hydrochloric acid (1.19) Nitric acid (1.40) Hydrofluoric acid (40%)	30 ml 15 ml 30 ml	Secs to mins. !!! See Appendix A.	Mo, W, V, Nb, Ta, and their alloys.
Cr M3	Dist. water Nitric acid (1.40) Hydrofluoric acid (40%)	75 ml 35 ml 15 ml	10–20 min. !!! See Appendix A.	Mo, W, and V.

Microetching

Preparation:

Grinding:

Same as for macroetching.

Polishing:

a. With diamond paste or alumina slurry down to finest particle size; best on wheel with hardwood cover. In most cases, attack-polishing with one of the electrolytes listed under (b) is of advantage.

b. Electrolytic:

	Electrolyte		Conditions	Remarks
1.	Methanol (95%) or dist. water Sulfuric acid (1.84)	875 ml 125 ml	30 s to 2 min, 6–18 V dc, stainless steel cathode.	Mo !!! See Appendix A.
2.	Sulfuric acid (1.84) Hydrofluoric acid (40%)	900 ml 100 ml	5–10 min, approx 40 °C (100 °F)	Nb, Ta. !!! See Appendix A.
3.	Dist. water Sodium hydroxide	1000 ml 100 ml	10–25 min, 6 V dc, graphite cathode.	W, Cr.
4.	Glacial acetic acid Perchloric acid (60%)	900–950 ml 50–100 ml	Secs to 2 min, 25–30 V dc. !!! See Appendix A.	Cr, V, Re.

c. Electrolytic lapping:			
1. Dist. water Sodium hydroxide	100 ml 2 g	1–10 min, 10–20 V dc, stainless steel cathode. Velvet cloth.	W.
2. Dist. water with alumina slurry No. 3 (0.05 μm, Linde B) Potassium ferricyanide	30 ml 1 g	5–10 min, 40 V dc, stainless steel cathode. Tough nylon cloth.	Mo.
3. Dist. water with alumina slurry No. 3 (0.05 μm, Linde B) Hydrogen peroxide (30%)	300 ml 10 ml	20 min, 1–10 mA/cm^2, stainless steel cathode.	Nb.
4. Methanol (95%) Nitric acid (1.40) Hydrofluoric acid (40%) Citric acid	510 ml 170 ml 50 ml 5 g	40 s, 5 mA/cm^2, stainless steel cathode. !!! See Appendix A.	Nb.
d. Chemical:			
Sulfuric acid (1.84) Nitric acid (1.40) Hydrofluoric acid (40%)	50 ml 20 ml 20 ml	5–10 s. !!! See Appendix A.	Preferentially for Ta and Nb.

Etching:

No.	Reagent		Conditions	Remarks
Cr m1	Nitric acid (1.40) Hydrochloric acid (1.19)	20 ml 60 ml	5–60 s. Do not store!	Cr and Cr-base alloys.
Cr m2	Nitric acid (1.40) Hydrofluoric acid (40%)	20 ml 60 ml	Approx 10 s. !!! See Appendix A.	Cr, Nb, and alloys.
Cr m3	a. Dist. water Potassium hydroxide b. Dist. water Potassium ferricyanide	100 ml 10 g 100 ml 10 g	15–60 s. Equal amounts of (a) and (b). Use fresh only! For Mo and W, sodium hydroxide and sodium ferricyanide may also be used.	Cr, Mo, Mo-Cr alloys (up to 80% Cr). Mo-Fe alloys. W and W-base alloys. Mo-Re alloys. Re and Re-base alloys.
Cr m4	Dist. water Nitric acid (1.40) Hydrofluoric acid (40%) (Concentration variable)	50 (50) ml 50 (25) ml 50 (5) ml	Secs to mins. !!! See Appendix A.	Ta, Nb, and their alloys. Cr and Cr silicide. Re silicide. W-Th alloys.

Cr m5	Dist. water Hydrogen peroxide (3%) Ammonia water	50 (70) ml 50 (20) ml 50 (10) ml	Secs to mins. 10 min, boiling.	Mo and Mo-Ni alloys. W and W alloys. Proportions in parentheses for Nb and its alloys.
Cr m6	Dist. water Hydrofluoric acid (40%) (Concentration variable)	50 ml 50 ml	10 s. !!! See Appendix A.	Ta and Ta-base alloys.
Cr m7	Hydrofluoric acid (40%) Nitric acid (1.40) Lactic acid (90%)	10 (10) ml 30 (10) ml 60 (30) ml	15–20 sec Do not store! !!! See Appendix A.	Nb, Ta, Mo and their alloys. Proportions in parentheses for Ta and Nb alloys.
Cr m8	Glacial acetic acid Nitric acid (1.40) Hydrofluoric acid (40%) (Concentration variable)	50 ml 20 ml 5 ml	10–30 s. !!! See Appendix A.	Ta-base alloys. Nb-base alloys.
Cr m9	Dist. water Hydrogen peroxide (30%) (Concentration variable)	100 ml 1 ml	30–90 s. Boiling.	W and W-base alloys.
Cr m10	Glycerol Hydrofluoric acid (40%) Nitric acid (1.40)	10–20 ml 10 ml 10 ml	Up to 5 min. !!! See Appendix A.	Mo, Ta, Nb, Mo-Ti alloys. Ta-Nb alloys. Pure V and V-base alloys.
Cr m11	Hydrochloric acid (1.19) Nitric acid (1.40) Glycerol	30 ml 15 ml 45 ml	Secs to mins.	Cr and Cr alloys.
Cr m12	Hydrochloric acid (1.19) Hydrogen peroxide (3%)	10 ml 10 ml	Secs to mins.	W-Co alloys containing 10–70% W.
Cr m13	Dist. water Picric acid Sodium hydroxide	100 ml 2 g 25 g	15 s. Boiling.	Eutectic W-Co alloys. W component turns black.
Cr m14	Dist. water Hydrofluoric acid (40%) Nitric acid (1.40) Sulfuric acid (1.84) (Concentration variable)	50 (0) ml 20 (20) ml 10 (20) ml 15 (50) ml	Secs to mins. !!! See Appendix A.	Ta and Ta-base alloys. Nb and Nb-base alloys. Nb-Cr alloys. Mo and Mo alloys.
Cr m15	Dist. water Sodium hydroxide	100 ml 10 g	Secs to mins.	Pure Ta.

Cr m16	*Electrolytic:* Dist. water Sodium hydroxide	 100 ml 10 g	Secs to mins, 1.5–6 V dc, stainless steel cathode.	W and W-base alloys. Ta and its alloys.
Cr m17	*Electrolytic:* Dist. water Oxalic acid	 100 ml 10 g	2–5 s, 6 V dc, stainless steel cathode. 1 min	Mo (grain contrast) Pure V and V-base alloys. Cr and Cr-base alloys. Re and Re-base alloys.
Cr m18	*Electrolytic:* Dist. water or ethanol (96%) Hydrochloric acid (1.19)	 95 ml 5 ml	Secs to mins, 5–10 V dc, stainless steel cathode.	V and V-base alloys. Alloys of high Cr content. Mo-Cr-Fe alloys. Mo and Mo alloys. U-Nb alloys. Ferrovanadium. Cr silicides. Galvanic Cr layers.
Cr m19	*Electrolytic:* Glacial acetic acid Perchloric acid (60%)	 95 ml 5 ml	15 s, 30–50 V dc, stainless steel or Pt cathode. !!! See Appendix A.	Cr and Cr-base alloys. Fe-Cr alloys. V.
Cr m20	*Electrolytic:* Methanol (95%) Sulfuric acid (1.84) Hydrofluoric acid (40%)	 100 ml 5 ml 1 ml	10–20 s, 50–60 V dc. !!! See Appendix A.	W and W-base alloys. Mo and Mo-base alloys.
Cr m21	*Electrolytic:* Methanol (95%) Sulfuric acid (1.84) Hydrochloric acid (1.19)	 75 ml 10 ml 25 ml	30 s, 30 V dc, stainless steel cathode. !!! See Appendix A.	Ta and Ta-base alloys. Mo and Mo-base alloys. V and V-base alloys.
Cr m22	*Electrolytic:* Dist. water Nitric acid (1.40) Hydrofluoric acid (40%)	 65 ml 17 ml 17 ml	Secs to mins, 12–30 V dc, Pt cathode. !!! See Appendix A.	Pure Nb. Mo.

Copper

Cu

Pure copper is very soft, and metallographic preparation often results in deformed layers. Grinding and polishing residues are easily embedded into the surface of the microsection. The preparation of copper alloys generally does not pose any difficulties.

Macroetching

Preparation:

Coarse grinding is sufficient (wet on silicon carbide papers down to 400 grit).

Etching:

No.	Etchant		Conditions	Remarks
Cu M1	Dist. water or ethanol (96%) Hydrochloric acid (1.19) Iron (III) chloride (Concentration variable)	120 (100) ml 30 (6) ml 10 (20) g	Few mins.	Cu and all types of brasses. Grain contrast. Bronzes. Al bronze. Formation of dendrites in alpha alloys.
Cu M2	a. Dist. water Mercury (II) nitrate b. Dist. water Nitric acid (1.40)	100 ml 1 g 100 ml 1 ml	Mix (a) and (b) in equal amounts. Time until sample begins to crack is an indication of stresses. !!! See Appendix A.	Verification of stresses in brass. Samples with stresses begin to crack after some time.
Cu M3	Dist. water Nitric acid (1.40)	90 ml 10–60 ml	Few mins.	Cu and all types of brasses. Grains and cracks.
Cu M4	Dist. water Nitric acid (1.40)	50 ml 50 ml	Few mins.	Cu and brasses. Grain contrast.
Cu M5	Dist. water Nitric acid (1.40) Silver nitrate	50 ml 50 ml 5 g	Secs to mins. Deep etching.	All types of Cu and Cu alloys.
Cu M6	Dist. water Nitric acid (1.40) Sulfuric acid (1.84) Ammonium chloride Chromium (VI) oxide	100 ml 50 ml 8 ml 7.5 g 40 g	Secs to mins. !!! See Appendix A.	Brasses containing Si. Si bronzes.
Cu M7	Dist. water Ammonium persulfate	100 ml 25 g	Secs to mins.	Brasses. Especially brasses containing Co.

Microetching

Preparation:

Grinding:

a. Wet on silicon carbide papers down to 600 grit.

b. Prepare by microtome using cutters of cemented carbides or, still better, diamond; eliminates polishing. For all copper alloys with a hardness of less than 150 HV.

Polishing:

a. With alumina slurry or diamond paste down to finest particle size. Possibly under addition of a solution of ammonium hydroxide or copper ammonium persulfate; this frequently eliminates etching.
b. Same as under (a) but additional etch-polishing sequence with etchant No. Cu m6.
c. Electrolytic:

1. Dist. water Phosphoric acid (1.71)	300 ml 700 ml	5–15 min, 1.5–2 V dc, Cu cathode.	For all types of Cu and Cu alloys, with exception of Sn bronzes. After polishing, rinse in a 20% aqueous solution of phosphoric acid.
2. Dist. water Phosphoric acid (1.71) Sulfuric acid (1.84)	300 ml 670 ml 100 ml	15 min, 2 V dc, Cu cathode.	Same as for (c. 1) above. Also used for Sn bronzes.
3. Dist. water Chromium (VI) oxide	100 ml 1 g	3–6 s, 6 V dc, Al cathode.	Al and Be bronzes.

d. Electrolytic lapping:

Dist. water Sodium thiosulfate Copper (II) nitrate Thiourea	885 ml 12 g 1 g 2 g	5–7 min, 5 mA/0.4 cm^2. 8–9 mA/0.64 cm^2. 10 mA/0.8 cm^2.	 High-purity Cu. Rolled Cu. 80Ag-20Cu.

e. Chemical:

1. Nitric acid (1.40) Hydrochloric acid (1.19) Phosphoric acid (1.71) Glacial acetic acid (Concentration variable)	30 ml 10 ml 10 ml 50 ml	1–2 min, up to 80 °C (180 °F).	All types of Cu and Cu alloys.
2. Dist. water Nitric acid (1.40) Chromium (VI) oxide (Concentration variable)	35 (100) ml 40 (7) ml 25 (27) g	Oxide layer formed during polishing is removed with 10% hydrofluoric acid. !!! See Appendix A.	Grain boundaries. Cu-Al alloys. Alpha, gamma phase.

Etching:

No.	Etchant		Procedure	Remarks
Cu m1	Dist. water or ethanol (95%) Hydrochloric acid (1.19) Iron (III) chloride (Concentration variable)	100–120 ml 20–50 ml 5–10 ml	Secs to mins. Possibly followed by Cu m4.	All types of Cu. Cu-Be alloys. Brasses (colors beta brass). Special bronzes. Al bronze with eutectoid. German silver.

Cu m2	Dist. water Ammonium persulfate (possibly) Hydrochloric acid (1.19)	100 ml 10 g 10 ml	Secs to mins. Possibly heated. Use fresh only! Mixing of Cu m1 and Cu m2 in equal amounts or 3:1, or cyclic immersion etching in Cu m1 and Cu m2.	Cu. Brasses. Bronzes. Al bronzes. Cu-Ni and Cu-Ag alloys. German silver. Grain contrast of alpha brasses. Cu welds. Also useful for macroetching.
	Ammonium persulfate	20 g		Verification of (100) rolling texture in Cu.
Cu m3	Dist. water Copper (II) ammonium chloride Add ammonia water; precipitate dissolves	120 ml 10 g	5–60 s. If dissolution too severe, dilute with dist. water.	Cu. Alpha-beta brass. Special brass. Al brass. Red-cast bronze. German silver. Cu-Sn alloys.
Cu m4	Dist. water Sulfuric acid (1.84) Sodium or potassium dichromate	80 ml 5 ml 10 g	3–30 s. Immediately before use, add two drops of hydrochloric acid (1.19)	Brasses. Alpha bronze. German silver. Cu-Be, Cu-Cr, Cu-Ni, Cu-Mn, Cu-Si alloys.
Cu m5	Sat. aqueous sol. of sodium thiosulfate Potassium metabisulfite	50 ml 5 g	6–8 min	Color and grain contrast etchant for Cu.
	Potassium metabisulfite (Klemm's reagent)	1 g	3 min	Alpha and beta brass.
Cu m6	Dist. water Ethanol (96%) Iron (III) nitrate	100 ml 100 ml 10 g	Attack-polishing without adding abrasive on velvet.	Fast, good polish and etch for Cu and Cu with oxide and sulfide inclusions.
Cu m7	Dist. water Hydrochloric acid (1.19) Nitric acid (1.40)	30 (100) ml 10 (8) ml 10 (25) ml	Secs to mins.	Multiple constituent Sn bronzes. Delta-phase. Proportions in parentheses for Cu-Ga alloys.
Cu m8	Dist. water Cold, sat. aqueous sol. of sodium thiosulfate Potassium metabisulfite	45 ml 5 ml 20 g	3–5 min.	Color etchant and grain contrast of bronzes.
Cu m9	Sat. aqueous sol. of chromium (VI) oxide		5–30 s. !!! See Appendix A.	Grain boundaries. Pure Cu. Brasses. Bronzes. German silver.

Cu m10	Dist. water Nitric acid (1.40) Hydrofluoric acid (40%)	100 ml 13 ml 6.6 ml	Few mins. !!! See Appendix A.	Al bronzes which are difficult to etch otherwise.
Cu m11	Dist. water Ammonia water Hydrogen peroxide (3%)	25 ml 25 ml 5–25 ml	Secs to mins. Use fresh only! Possibly addition of 1–5 ml solution of potassium hydroxide (20%).	Most types of Cu and Cu alloys. Cu-Ag solder layers. Mn, P, Be, Al-Si bronzes. Alpha brasses.
	Little hydrogen peroxide			Grain boundaries.
	More hydrogen peroxide			Grain contrast.
Cu m12	Dist. water Nitric acid (1.40) (Concentration variable)	50 ml 50 ml	Secs to mins.	Most types of Cu and Cu alloys. Flow lines in brass.
Cu m13	*Electrolytic:* Aqueous sol. of chromium (VI) oxide (1%)		3–6 s, 6 V dc, Al cathode.	Al bronzes. Cu-Be alloys.
Cu m14	*Electrolytic:* Dist. water Sulfuric acid (1.84) Sodium hydroxide Iron (II) sulfate	 950 ml 50 ml 2 g 15 g	Up to 15 s, 8–10 V dc, Cu cathode.	Beta brass. German silver. Monel and Cu-Ni alloys. Bronzes.
Cu m15	*Electrolytic:* Dist. water Phosphoric acid (1.71)	 90 ml 10 ml	5–10 s, 1–8 V dc, Cu cathode.	All types of Cu. Cartridge brass. Tombac. Muntz metal. Easily machinable brasses.
Cu m16	*Electrolytic:* Dist. water Phosphoric acid (1.71) Hydrofluoric acid (40%) Chromium (VI) oxide	 480 ml 960 ml 15 ml 9 g	30 s, 1.1 A/cm^2, 5 V dc, stainless steel cathode. !!! See Appendix A.	Brasses.

Iron, Steel, Cast Iron

Fe

The preparation of iron, steel, and ferrous alloys usually does not give rise to particular difficulties. However, the pronounced multiplicity of properties and microstructures in this category requires differing methods of preparation. Unalloyed irons and low-carbon steels are easily deformable, and therefore deformation layers frequently form. Steels, as well as cast irons (2-4.5% C) vary from tough to brittle. Hardened steels should not be heated during preparation since this might cause changes in microstructure. In addition, in materials that have undergone a forming process, the direction of deformation should be taken into consideration. Fiber orientation, grain elongation, banded structure, and nonmetallic inclusions are better judged in longitudinal sections.

Macroetching

Preparation:

The surface quality resulting from machining, coarse grinding, or sectioning is often sufficient. Usually, wet grinding on silicon carbide papers down to 400 grit is satisfactory. In special cases, for verification of segregation and flow lines, wet grinding down to 600 grit or lapping with silicon carbide powders from 320 to 800 grit is required. When examining surface imperfections, etching is done directly.

Etching:

No.	Etchant		Conditions	Remarks
Fe M1	Ethanol (96%) Nitric acid (1.40) (Concentration variable)	90 ml 10 ml	1–5 min. Deep etching.	Versatile, useful for Fe and steel. Carburized, or decarburized zones. Segregation. Also useful for microetching.
Fe M2	Dist. water Hydrochloric acid (1.19) (Concentration variable)	50 ml 50 ml	10–30 min, up to 80 °C (180 °F). Remove film under running water. For corrosion-resistant steels, immerse in warm, 20% aqueous sol. of nitric acid (1.40).	Versatile etchant for alloyed and unalloyed steels. Deep etchant for surface control and segregation. Porosity, hardness indentations, fractures, inclusions, dendrites, flow lines, ferrite.
Fe M3	Hydrochloric acid (1.19) Sat. aqueous sol. of copper (II) sulfate (Marble's reagent)	50 ml 25 ml	Secs to mins.	Austenitic and high-temperature steels. Fe-Cr-Ni cast alloys. Depth of nitriding.
Fe M4	Dist. water Hydrochloric acid (1.19) Copper (II) chloride (Fry's reagent)	100 ml 120–180 ml 45–90 g	5–20 min. Prior to etching, heat at 150–200 °C (300–400 °F) for 5–30 min. After etching, immerse in conc. hydrochloric acid, rinse in water, and neutralize in ammonia water.	Flow lines in low-carbon N_2 steels. Thomas steel. Compare microetchant Fe m13.

Fe M5	Dist. water Sulfuric acid (1.84) (Sulfur-printing test) (Baumann print)	100 ml 5 ml	Silver bromide paper is saturated with solution, and firmly pressed against the prepared surface. After 1–5 min, rinse, fix (6 g sodium thiosulfate in 100 ml water), wash and dry.	Verification, arrangement, and distribution of Fe and Mn sulfide inclusions. Sulfide reacts with sulfuric acid, forming hydrogen sulfide which combines to darkish silver sulfide with the silver from the silver bromide in the photographic paper.
Fe M6	Dist. water Ethanol (96%) Hydrochloric acid (1.19) Iron (III) chloride Tin (II) chloride Add hydrochloric acid last (Oberhoffer's reagent)	500 ml 500 ml 42 ml 30 g 0.5 g	Secs to mins. Microsection must be polished. After etching, rinse in a 4:1 mixture of ethanol and hydrochloric acid.	Steels and special steels. Fe-enriched areas appear dark. Blowhole segregation. Primary structure. Fiber orientation.
Fe M7	Dist. water Copper (II) ammonium chloride (Heyn's reagent)	120 ml 20 g	2–10 min. Cu precipitates wiped with water.	Phosphorus segregation in low-carbon steels. Fiber orientation. Welding zones. Grain contrast. Primary structure.
Fe M8	Sat. aqueous sol. of sodium thiosulfate Potassium metabisulfite (Concentration variable) (Klemm's reagent)	50 ml 1 g	Secs to mins.	Phosphorus distribution in cast steel and cast iron.

Microetchant Fe m22 is also used for macroetching.

Microetching

Preparation:

Compare suggestions for macroetching.

Grinding:

Wet on silicon carbide papers down to 600 grit. For hardened steels, avoid heating.

Polishing:

a. Alumina slurry on felt (billiard cloth or wool). Toward the end of the polishing cycle, add a lot of water (possibly 0.1% aqueous solution of ammonium tartrate) to wash off adhering alumina particles.

b. Diamond paste, particle size 6-1 μm; 0.25 μm only for extreme requirements in surface quality (danger of overpolishing). Diamond polishing is particularly suitable for heterogeneous materials with microstructural components of differing hardness such as cast iron, malleable cast iron, steels containing carbides. Diamond polishing helps avoid relief formation. Graphite formation and nonmetallic inclusions can be studied in the unetched state. Try to avoid pull-outs!

c. Electrolytic:

1. Dist. water Sulfuric acid (1.84)	400 ml 600 ml	2–6 min, 1.5–6 V dc, Ni cathode.	Corrosion-resistant steels. Pure iron.
2. Dist. water Chromium (VI) oxide	830 ml 620 g	2–10 min, 1.5–9 V dc, Ni cathode. !!! See Appendix A.	Corrosion-resistant steels. High-alloy steels.

Etching:

Pure iron: Fe m1, 3, 4, 9, 19, 20, 21
Structural steels and heat treated steels: Fe m1–9, 12, 13, 14, 16, 17, 19, 20, 21, 24
High-temperature steels: Fe m1–12, 17, 18, 22–25
Tool steels: Fe m1–10, 14, 15, 18
Stainless steels: Fe m5, 6, 10, 11, 12, 21–25

No.	Etchant		Conditions	Remarks
Fe m1	Ethanol or methanol (95%) Nitric acid (1.40) (Nital) See note at Fe m3.	100 ml 1–10 ml	Secs to mins. Caution: not to exceed 10% solution. Explosive!	Most common etchant for pure iron, low-carbon steels, alloy steels, and gray cast iron. Segregations can cause uneven attack.
Fe m2	Ethanol or methanol (95%) Hydrochloric acid (1.19) (Concentration variable)	100 ml 20 ml	5–30 min. Use fresh only! Possible addition of hydrogen peroxide (30%).	Differentiation of cubic and tetragonal martensite. Ni-containing Cr steels. High-temperature steels.
Fe m3	Ethanol (96%) Picric acid (Concentration variable) (Picral) Nital and Picral (Fe m1 and Fe m3) may be mixed 1:1	100 ml 2–4 g	Secs to mins. !!! See Appendix A.	Generally used for iron and heat treated steels. Pearlite, martensite, and bainite. Lower contrast than Fe m1. Uniform etching, even with segregations. Fe_3C stained light yellow.
Fe m4	Methanol or ethanol (96%) Nitric acid (1.40) Picric acid	100 ml 0.2 ml 0.3 ml	Secs to mins.	Compare Fe m1 and m3. Improvement of contrast over Fe m3. Sometimes prevents uneven attack of Fe m1.
Fe m5	Ethanol (96%) Hydrochloric acid (1.19) Picric acid Possibly a few drops hydrogen peroxide (3%) or wetting agent	100 ml 1–5 ml 1–4 g	Secs to mins. To bring out grain boundaries, heat for 10 min at 300–500 °C (570–930 °F) before etching. !!! See Appendix A.	Increases attack of Fe m2 in alloyed steels. Grain-boundary precipitates in Cr-Ni steels become visible.

Fe m6	Ethanol or methanol (95%) Hydrochloric acid (1.19) Nitric acid (1.40)	85 ml 1–10 ml 1–5 ml	Up to several mins. !!! See Appendix A.	Grain boundaries in heat treated tool steels. Alloyed Cr steels.
Fe m7	Ethanol (96%) Nitric acid (1.40) Hydrochloric acid (1.19) Picric acid Possible addition of wetting agent	80 ml 10 ml 10 ml 1 g	Secs to mins. !!! See Appendix A.	Grain boundaries in martensitic microstructures.
Fe m8	Dist. water Sodium hydroxide Picric acid	75 ml 25 g 2 g	3–15 min, 50 °C (120 °F). !!! See Appendix A.	Cementite (Fe_3C) with up to 10% Cr is stained dark. For more than 10% Cr, no staining. $(Fe,Cr)_7C_3$, $(Fe,Cr)_{23}C_6$, WC and VC are not stained.
Fe m9	Dist. water Potassium or sodium hydroxide Potassium ferricyanide (Concentration variable) (Murakami's reagent)	100 ml 10 g 10 g	2–20 min, 20–50 °C (70–120 °F). Use fresh only! !!! See Appendix A.	Fe_3C with more than 10% Cr is stained dark more rapidly than Fe_3C with less Cr. $(Fe,Cr)_7C_3$, $(Fe,Cr)_{23}C_6$, and iron phosphide are stained.
Fe m10	Dist. water Hydrochloric acid (1.19) Nitric acid (1.40) Vogel's special reagent (V2A etchant)	100 ml 100 ml 10 ml 0.3 ml	Secs to mins, room temperature to 50 °C (120 °F).	Mn-alloyed Cr-Ni steels. Sigma phase and ferrite. Fine microstructures in alloyed steels.
Fe m11	Glycerol Nitric acid (1.40) Hydrochloric acid (1.19) (Concentration variable) (Vilella's reagent)	45 ml 15 ml 30 ml	Secs to mins. !!! See Appendix A.	Stainless steels with high Cr content. Cr-Ni cast steels.
Fe m12	Glycerol Nitric acid (1.40) Hydrofluoric acid (40%) (Concentration variable)	20–40 ml 10 ml 20 ml	Secs to mins. !!! See Appendix A.	Steels with high Si content.

Fe m13	Dist. water Ethanol or menthanol (95%) Hydrochloric acid (1.19) Copper (II) chloride (Fry's reagent)	30 ml 25 ml 40 ml 5 g	Secs to mins. Prior to etching, heat to 150–200 °C (300–400 °F). Compare macroetchant Fe M4.	In normalized, low-carbon steels containing nitrogen, deformed areas can be distinguished next to undeformed areas. Flow lines in forgings.
Fe m14	Sat. sol. of aqueous picric acid, possibly diluted Copper (II) chloride Wetting agent	100 ml 80 mg 60 ml	30–60 s, 75–85 °C (160–185 °F). Wipe black film on specimen with aqueous sol. of ammonium hydroxide on cotton	Austenite grain boundaries in tempered steels and case-hardened steels.
Fe m15	Same as Fe m14 plus Hydrochloric acid (1.19)	1 ml	Secs to mins.	Austenite grain boundaries in cold work steels.
Fe m16	Dist. water Ethanol (96%) Saturated picric acid (in alcohol) Wetting agent Iron (II) chloride Zinc (II) chloride	25 ml 75 ml 20 ml 14 ml 10 g 3 g	Secs to mins.	Austenite grain boundaries in case-hardened steels.
Fe m17	Ethanol (96%) Wetting agent Ammonia water Hydrochloric acid (1.19) Picric acid Copper (II) ammonium chloride	50 ml 10 ml 1 ml 1 ml 3 g 1 g	Secs to mins.	Austenite grain boundaries in valve steels, tempered steels, case-hardened steels.
Fe m18	Ethanol (96%) Wetting agent Picric acid Copper (II) ammonium chloride	48 ml 10 ml 6 g 1 g	Secs to mins.	Austenite grain boundaries in cold work steels.
Fe m19	Dist. water Potassium hydroxide Potassium ferricyanide	60 ml 30 g 30 g	20–40 s, 50 °C (120 °F). Use fresh only! !!! See Appendix A.	Sigma phase is stained. Austenite remains unchanged. Ferrite turns yellow-brown.
Fe m20	Dist. water Sodium thiosulfate Citric acid Cadmium chloride (Beraha's reagent)	100 ml 24 g 3 g 2–2.5 g	20–40 s. Dissolve chemicals in given sequence. Each constituent must be completely dissolved before the next one is added. Store in dark bottle, max 20 °C (70 °F). Filter solution before use. Remains useful for 4 h.	Tint etchant for pure iron and carbon steels, cast iron, and alloyed steels. Differentiation of ferrite (brown to violet), carbides, phosphides, and nitrides (light).

Fe m21 Dist. water Potassium meta-bisulfite	100 ml 10 g	Secs to mins. Possible pre-etching in Fe m1	Pure iron, carbon steels, low-alloy steels. Tinting of ferrite, martensite, bainite, and sorbite. Etch pits, subgrain boundaries, twins. Carbides, nitrides, phosphides, and sulfides are not attacked.
Fe m22 Dist. water Ammonium persulfate Used also as a macroetchant.	100 ml 10 g	5–20 s. Use fresh only!	Grain contrast of ferrite in low-carbon steels. Transformer sheet metals, austenitic Ni-steels.
		Approx 2 min	Carbides and phosphides in low-alloy steels. Only matrix is attacked.
		Electrolytic: Secs to mins, 6 V dc, Ni cathode.	High-alloy Cr-Ni steels. (Transformation and precipitation microstructures, carbides). Hardened steels with martensite.
Fe m23 *Electrolytic:* Dist. water Lead acetate	 100 ml 10 g	Secs, 2 V dc, stainless steel cathode.	Fe-Cr-Ni cast alloy. Sigma phase (blue-red), ferrite (dark blue), austenite (light blue), and carbide (yellow) in stainless and high-alloy heat-resistant steels.
Fe m24 *Electrolytic:* Ammonium hydroxide conc.		30–60 s, 1.5–6 V dc, Pt cathode.	High-alloy steels. Etches carbides only. Leaves sigma phase unchanged.
Fe m25 *Electrolytic:* Dist. water Chromium (VI) oxide	 100 ml 10 g	3–60 s, 3–6 V dc, Pt cathode. !!! See Appendix A.	Cr and Cr-Ni steels. Cementite is attacked rapidly, austenite less rapidly, ferrite and phosphide are attacked least. Carbide etchant. Corrosion-resistant malleable steel.
		Immersion etching	Cast iron.

No.	Etchant		Conditions	Remarks
Fe m26	*Electrolytic:* Dist. water Sodium hydroxide	100 ml 40 g	5–60 s, 1–3 V dc, Pt cathode.	Verification of sigma phase. First, stains yellow to dark brown. Then, staining of ferrite commences. Continued etching develops carbides.
			Potentiostatic: at +450 mV	More uniform, better quality etching.
			at −700 mV	Staining of cementite.

Ge

Se

Si

Te

Germanium, Selenium, Silicon, Tellurium, A_{III} B_V and A_{II} B_{VI} Alloys

Selenium and its compounds are toxic. (!!! See Appendix A and references on safety and toxicology in Appendix C.)

Microetching

Preparation:

Grinding:

Lapping with silicon carbide, boron carbide, or aluminum oxide slurries in water (grit size 320–800).

Polishing:

a. Use diamond paste or alumina slurries down to finest particle size. Etch-polishing sequence with solutions used for chemical polishing (c. 1) or (c. 2).

b. Electrolytic:

Methanol (95%) Butyl glycol Perchloric acid (65%)	590 ml 350 ml 60 ml	30–60 s, 25–35 V dc, stainless steel cathode. !!! See Appendix A.	

c. Chemical:

1. Hydrofluoric acid (40%) Nitric acid (1.40) Glacial acetic acid Bromine	15 ml 25 ml 15 ml 3–4 drops	5–10 s. !!! See Appendix A.	Ge.
2. Nitric acid (1.40) Hydrofluoric acid (40%)	20 ml 5 ml	5–10 s !!! See Appendix A.	Si.

Etching:

No.	Etchant		Conditions	Remarks
Ge m1	Dist. water Nitric acid (1.40) Hydrofluoric acid (40%)	5 (90) ml 25 (5) ml 25 (5) ml	5–20 s. !!! See Appendix A.	Si, Ge, and their alloys. InSb. Proportions in parentheses especially for Si. Also used for chemical polishing at approx 50 °C (120 °F)

Ge m2	Ethanol (96%) Hydrochloric acid (1.19) Picric acid	100 ml 5 ml 1 g	Secs to mins.	Ge-In alloys with Ag, Au, Bi, Cu additions. Grain boundaries.
Ge m3	Hydrofluoric acid (40%) Nitric acid (1.40)	10 ml 10 ml	Secs to mins. !!! See Appendix A.	Si, Ge, and their alloys. InSb. Etch pits on (111) planes, p-n junctions.
Ge m4	Dist. water Hydrofluoric acid (40%) Hydrogen peroxide (30%)	40 ml 10 ml 10 ml	1–3 min. !!! See Appendix A.	Se, Ge, and their alloys. InAs, InSb, GaSb, GaAs, AlSb, ZnTe, CdTe, InP.
Ge m5	Nitric acid (1.40) Hydrofluoric acid (40%) Glacial acetic acid Bromine (CP4 solution) Without bromine	50 ml 30 ml 30 ml 0.6 ml	3–25 s. Mix 30 min before use. !!! See Appendix A.	Si, Ge, and their alloys. GaSb. InSb. Dislocations on (100) and (111) planes of Si. InAs.
Ge m6	Nitric acid (1.40) Possibly diluted with dist. water, and/or mixed with hydrochloric acid (1.19)		Secs to mins.	Ge, Te, Se. Tellurides. Selenides and Zr silicide.
Ge m7	Dist. water Hydrochloric acid (1.19) Iron (III) chloride	50 ml 50 ml 20 g	Secs to mins. Boiling.	Ge and its alloys. Grain contrast. Also suitable as a macroetchant.
Ge m8	Dist. water Hydrofluoric acid (40%) Nitric acid (1.40) Silver nitrate	40 ml 40 ml 20 ml 2 g	30 s to 2 min !!! See Appendix A.	Ge and its alloys. GaAs, InAs, AlAs. Grain boundaries. Dislocations on (111) planes.
Ge m9	Dist. water Sodium hydroxide	100 ml 50–100 g	Secs to mins.	Si, Te, Se.
Ge m10	*Electrolytic:* Dist. water Oxalic acid	100 ml 10 g	10–20 s, 4–6 V dc, stainless steel cathode.	Ge and its alloys. Grain boundaries.

Hf
Zr

Hafnium and Zirconium

Pure hafnium and zirconium are soft, and metallographic preparation often results in deformation layers. Their alloys usually do not pose any particular difficulties.

Macroetching

Preparation:

Wet grinding on silicon carbide papers down to 600 grit. Possible coating of grinding paper with wax.

Etching:

No.	Etchant		Conditions	Remarks
Hf M1	Dist. water Nitric acid (1.40) Hydrofluoric acid (40%)	70 ml 30 ml 5 ml	Secs to mins. Swab. !!! See Appendix A.	Zircaloy-2 and Hf.
Hf M2	Dist. water Nitric acid (1.40) Hydrofluoric acid (40%)	45 ml 45 ml 10 ml	Secs to mins. Swab. !!! See Appendix A.	Zr and Hf and their alloys with high additional alloying constituents. With smaller amounts of additional alloying constituents, use ethanol (96%) instead of water.
Hf M3	Hydrogen peroxide (30%) Nitric acid (1.40) Hydrofluoric acid (40%)	45 ml 45 ml 10 ml	Secs to mins. Swab. !!! See Appendix A.	Zr and Hf and their alloys with small additional constituents

Microetching

Preparation:

Grinding:

Same as under macroetching. Also, prepare by microtome using cutters of cemented carbide; diamond cutters tend to roughen the surface.

Polishing:

a. With diamond paste and alumina slurry down to finest particle size. Possible etch-polish sequence with one of the microetchants.

b. Electrolytic:

1. Glacial acetic acid Perchloric acid (70%)	90 ml 10 ml	30 s to 8 min, 12–18 V dc, stainless steel cathode. !!! See Appendix A.
2. Glycerol Hydrofluoric acid (40%) Nitric acid (1.40)	870 ml 43 ml 87 ml	1–10 min, 9–12 V dc, stainless steel cathode. Use fresh only! Do not store! !!! See Appendix A.

c. Chemical:

1. Dist. water or hydrogen peroxide (30%)	45 ml	5–10 s repeatedly. !!! See Appendix A.
Hydrofluoric acid (40%)	10 ml	
Nitric acid (1.40)	45 ml	

Etching:

Hafnium and zirconium are suitable for investigations under polarized light.

No.	Etchant		Conditions	Remarks
Hf m1	Nitric acid (1.40) Hydrofluoric acid (40%) Nitric acid (1.19)	15 ml 30 ml 30 ml	3–10 s, swab. 2-min immersion etch. !!! See Appendix A.	Zr and Hf-base alloys Zr-Nb alloys.
Hf m2	Dist. water Nitric acid (1.40) Hydrofluoric acid (40%)	45 ml 45 ml 10 ml	5–20 s. !!! See Appendix A.	Pure and low-alloy Hf and Zr for observations in polarized light. Zr-U, Zr-Al, Hf-Re alloys.
Hf m3	Dist. water Hydrofluoric acid (40%) Aqueous sol. of silver nitrate (5%) with dist. water	200 ml 5 ml 2 ml 100 ml (approx)	5–60 s. !!! See Appendix A. 10–60 s.	Hf-base alloys. Zr-base alloys.
Hf m4	Dist. water Hydrofluoric acid	100 ml 5–10 ml	1–5 s. !!! See Appendix A.	Zr, Zr-Be, Zr-H, and Zr-Nb alloys. Zircaloy.
Hf m5	Nitric acid Hydrofluoric acid (40%) (Concentration variable)	90 ml 10 ml	Secs to mins. !!! See Appendix A. With equal parts of the constituents	Zr-Th, Zr-Sn, Zr-Nb, Zr-Cu, Zr-Si, Zr-Ni alloys. Zr alloys with Al, Be, Fe, Ni, Si.
Hf m6	Glycerol Nitric acid (1.40) Hydrofluoric acid (40%) (Concentration variable)	85 (45) ml 10 (45) ml 5 (10) ml	Secs to mins. Do not store! !!! See Appendix A.	Zr-Mg, Zr-Mo, Zr-Sn, Zr-U, Zr-B, Zr-Fe, Zr-Ni alloys. Proportions in parentheses for alloys with low additions.
Hf m7	*Electrolytic:* Glycerol Nitric acid (1.40) Hydrofluoric acid (40%)	 100 ml 10 ml 5 ml	1–10 min, 9–12 V dc, Pt cathode. Do not store! !!! See Appendix A.	Zr and its alloys.

Mercury Alloys (Amalgamates)

Hg

Amalgamates are heat sensitive. They have a tendency to smear. Use only very light pressure, with external cooling of specimen. Microscopy should immediately follow preparation. Amalgamates are toxic (!!! See Appendix A.)

Microetching

Preparation:

Grinding:

Wet on already used silicon carbide paper (possibly coated with wax). Grit size 320–600; subsequently, 600-grit soft, already used.

Polishing:

Diamond paste (particle size 3-0.25 μm, approx 5–15 min). Subsequently, alumina slurry No. 3 (0.05 μm, Linde B)—or, even better, magnesia slurry on velvet or soft leather. Use light pressure, fast movement. Rinse in distilled water. Dry by pressing against blotting paper. Cool-air stream (hair drier) can also be used. Possible etch-polishing sequence. Grinding and polishing can be replaced by preparation with microtome.

Etching:

No.	Etchant		Condition	Remarks
Hg m1	a. Dist. water or ethanol (96%)	90 ml	1–3 min.	For most amalgamates.
	Hydrochloric acid (1.19)	10 ml		
	b. Dist. water	90 ml	Dip briefly.	Usually, (a) is sufficient.
	Aqueous sol. of iron (III) chloride (5%)	10 ml		Double etching in (a) and (b) results in greater contrast.
	Hydrochloric acid 1.19)	1 ml		
Hg m2	Glacial acetic acid		Approx 15 min	Hg-Sn and Hg-Sn-Cu alloys.
Hg m3	Ammonia water	100 ml	Secs to mins	Ag-Sn-Hg alloys.
	Hydrogen peroxide (30%)	25 ml		
	(possibly dist. water)	(100 ml)		

Etching solutions for the respective alloying metals of mercury are suitable in diluted form for amalgamates.

Mg

Magnesium

Deformation layers frequently form on magnesium. Also, there is a tendency toward mechanical twinning at relatively low deformation. Abrasives and polishing compounds are embedded into the surface. Some phases containing magnesium are attacked by water. Therefore, extended treatment involving water should be avoided. Magnesium dust and shavings are pyrophoric (!!! See Appendix A).

Macroetching

Preparation:

Coarse grinding on silicon carbide paper is usually sufficient. If water-sensitive components are present, grinding has to be carried out dry, or by using alcohol or kerosine as a lubricant with proper ventilation precautions.

Etching:

No.	Etchant		Conditions	Remarks
Mg M1	Dist. water Sol. of picric acid (64%) in ethanol (96%) Glacial acetic acid Possibly higher water content	20 ml 50 ml 20 ml	30s to 3 min Precipitate is washed off in hot water. !!! See Appendix A.	Grain size and flow lines in cast and forged parts.
Mg M2	Dist. water Acetic acid	100 ml 10 ml	30 s to 3 min Swab	Internal defects in cast ingots. Segregations. Flow lines in forged parts.
Mg M3	Dist. water Nitric acid (1.40)	100 ml 20 ml	30 s to 5 min	Internal defects in cast ingots. Segregations. Flow lines in forged parts. Mg-Mn and Mg-Zr alloys

Microetching

Preparation:

Grinding:

a. Observe instructions for macroetching.

b. Wet with distilled water on silicon carbide papers down to grit size 600 (for specimens which are insensitive to water).

c. Prepare by microtome using cemented carbide blade or, even better, diamond blade; eliminates polishing.

Polishing:

a. With alumina slurry on Microcloth, or with diamond (particle size 6-1 μm). Subsequently, on soft cloth with a suspension of 120 ml hot distilled water, 20 ml of aqueous solution of ammonium tartrate (5%), and 1 g MgO. Suspension should be filtered before use (through a fine cotton or nylon cloth).

b. Electrolytic:

Dist. water	250 ml	2 min, 25–50 °C (75–120 °F), 10 V dc, Mg or Al cathode.	
Ethanol (96%)	300 ml		
Phosphoric acid (1.71)	400 ml	!!! See Appendix A.	

c. Chemical

Nitric acid (1.40)		Imerse for 3-s periods. Polishing effect after about 1 min. Then etching commences.	Pure Mg.

Etching:

Magnesium is suitable for observations in polarized light.

No.	Etchant		Conditions	Remarks
Mg m1	Ethanol, methanol (95%), or dist. water Nitric acid (1.40)	100 ml 1–8 ml	Secs to mins.	Pure Mg and most Mg alloys, also in cast and forged state.

	Composition	Amount	Conditions	Use
Mg m2	Dist. water Oxalic acid	100 ml 2 g	6–10 s	Pure Mg, Mg-Mn, Mg-Al, Mg-Al-Zr, Mg-Th-Zr, Mg-Zn-Zr alloys. Also extruded types.
Mg m3	Dist. water Tartaric acid	90 ml 2–10 g	10 s to 2 min Polish dry only! Intense etch rinsing. Subsequently, rinse in alcohol. Dry in cold air stream.	Flow lines. Grain size in cast material. Mg-Al, Mg-Mn, and Mg-Mn-Al-Zn alloys.
Mg m4	Dist. water Ethylene glycol (1.11) Glacial acetic acid Nitric acid (1.40)	19 ml 60 ml 20 ml 1 ml	1–30 s. Swab. Rinse hot.	Most types of Mg and its alloys. Also castings and forgings.
Mg m5	Dist. water Ethylene glycol (1.11) Nitric acid (1.40)	24 ml 75 ml 1 ml	30–60 s.	All types of Mg alloys in cast or heat treated state. Grain-boundary etch.
Mg m6	Dist. water Hydrofluoric acid (40%) Possibly less hydrofluoric acid	90 ml 10 ml	3–30 s. !!! See Appendix A.	Technically pure Mg, Mg-Al-Zn, Mg-Zn-Th-Zr, Mg-Rare Earth-Zr alloys.
Mg m7	Ethanol (96%) Citric acid	100 ml 2–11 g	30 s	Mg and Mg-Cu alloys. Mg die-casting alloys.
Mg m8	Dist. water Nitric acid (1.40) Chromium (VI) oxide	85 ml 15 ml 12 g	10–30 s. Increase water content for alloys with high Al content. !!! See Appendix A.	Mg-Al alloys. Grain contrast in heat treated castings. Flow lines in forgings.
Mg m9	*Electrolytic:* Dist. water Sodium hydroxide	 100 ml 10 g	2–4 min, 4 V dc, Cu cathode. Etch immediately after polishing.	Complex Mg alloys containing Al, Zn, Cd, Bi.
Mg m10	*Electrolytic:* Dist. water Ethanol (96%) Phosphoric acid (1.71)	 20 ml 20 ml 40 ml	1–10 min, 10–35 V dc, Mg cathode. !!! See Appendix A.	Mg and Mg-base alloys.

Manganese

Mn

Microetching

Preparation:

Grinding:

Wet on silicon carbide papers down to 600 grit.

Polishing:

a. With diamond paste (particle size 6–0.25 μm), or alumina slurry on Microcloth.

b. Electrolytic:

Ethanol (96%)	50 ml	18 V dc, stainless	Mn. Mn-Cu alloys.
Glycerol	25 ml	steel cathode.	
Phosphoric acid (1.71)	25 ml	!!! See Appendix A.	

Etching:

No.	Etchant		Conditions	Remarks
Mn m1	Dist. water Hydrofluoric acid (40%)	90 ml 10 ml	Secs to mins. !!! See Appendix A.	Mn-Si-Ca alloys. Ferromanganese.
Mn m2	Ethanol (96%) Nitric acid (1.40) (Nital)	98 ml 2 ml	Secs to mins.	Mn-Fe, Mn-Ni, Mn-Cu, Mn-Co alloys.
Mn m3	Glycerol Hydrofluoric acid (40%) Hydrochloric acid (1.19) Nitric acid (1.40)	40 ml 30 ml 25 ml 10 ml	1–3 s. !!! See Appendix A.	Mn-Ge, Mn-Si, Mn-Sn-Ge, Mn-Sn-Si alloys.
Mn	Acetelyne acetone Nitric acid (1.40)	200 ml 1–2 ml	2–18 min. If possible, ultrasonically, to destroy oxide film.	Pure Mn. Low-alloy Mn containing Ni, Co, Fe, Ge, and Cu.

Nickel

Ni

Pure nickel is tough, and deformation layers frequently form. Nickel alloys do not pose any particular difficulties. Skin and respiratory poisoning may result from contact with nickel.

Macroetching

Preparation:

Wet grinding on silicon carbide papers down to 400 grit is normally sufficient.

Etching:

No.	Etchant		Conditions	Remarks
Ni M1	Dist. water Nitric acid (1.40) (Concentration variable)	50 ml 50–100 ml	20–30 min.	High-alloy Ni. Ni-Cu alloys (Monel). Ni-Al, Ni-Fe alloys. Porosity. Flow lines.
Ni M2	Dist. water Ethanol (96%) Hydrochloric acid (1.19) Copper (II) sulfate	50 ml 50 ml 50 ml 10 g	Secs to mins.	Ni and Ni-base alloys. Ni-Cu alloys. Ni-Cr-Fe alloys. Grain size in superalloys.
Ni M3	Dist. water Nitric acid (1.40) Copper (II) sulfate	10 (100) ml 20 (100) ml 10 (20) g	20–30 min.	Low-alloy Ni. Cracks. Porosity.
Ni M4	Glacial acetic acid Nitric acid (1.40) Hydrochloric acid (1.19)	50 ml 50 ml (75) ml	Secs to mins. Swab. Hydrochloric acid may possibly be omitted.	Ni alloys containing Cr and Fe. Welding joints.
Ni M5	Dist. water Nitric acid (1.40) Hydrochloric acid (1.19) Hydrogen peroxide (30%) (Concentration variable)	20–30 ml 0–20 ml 20 ml 10 ml	2 min. Use fresh only!	Inconel-type alloys on Ni-Cr and Ni-Fe-Cr basis. Ni-Nb, Ni-Ta, Ni-Si, Ni-Co-Cr alloys. High abrasion in superalloys.
Ni M6	Aqueous sol. of iron (III) chloride Hydrochloric acid (1.19) Nitric acid (1.40)	125 ml 600 ml 18.5 ml	5–10 min. Boiling.	Superalloys.

Microetching

Preparation:

Grinding:

a. Wet on silicon carbide papers down to 600 grit.
b. Prepare by microtome using cemented carbide blade or diamond blade; eliminates polishing. Hardness of specimen should not be greater than 150 HV.

Polishing:

a. Alumina slurry No. 1 (5 μm) and No. 1 C (1 μm; Linde C) on felt. Toward the end of the polishing cycle, add water to remove adhering abrasive particles.
b. Diamond paste, particle size 6 to 1 μm. For highest quality, also 0.25 μm. Etch-polishing sequence with solutions (c. 2) and (e) is of advantage for removal of deformed layer.
c. Electrolytic:
   1. Glacial acetic acid 700 ml, Perchloric acid (60%) 300 ml — 1–5 min, 40–100 V dc, Ni cathode.
      !!! See Appendix A.

2. Dist. water Sulfuric acid (1.84)	400 ml 600 ml	2–6 min, 1.5–6 V dc, Ni cathode.
3. Methanol (95%) Nitric acid (1.40)	660 ml 330 ml	10–60 s, 40–70 V dc, Ni cathode. !!! See Appendix A.

d. Electrolytic lapping:

1. Dist. water Potassium thiocyanate Disodium hydrogen phosphate Sodium fluoride	840 ml 50 g 50 g 60 g	Approx 5 min, 1.3 A/0.4 cm^2, ac	
2. Dist. water Sodium thiosulfate Copper (II) nitrate Thiourea	985 ml 12 g 1 g 2 g	3–4 min, 4–6 mA/0.7 cm^2. Can be used with and without the addition of alumina abrasive.	Cast Ni.
3. Dist. water Alumina slurry No. 3 (0.05 μm; Linde B) Ammonium chloride Potassium nitrate	100 ml 10 ml 5 g 2 g	0.5–1 min, 0.05–0.1 mA/0.08 cm^2, NiCr10 cathode.	

e. Chemical

Nitric acid (1.40) Sulfuric acid (1.84) Phosphoric acid (1.71) Glacial acetic acid	30 ml 10 ml 10 ml 50 ml	30–60 s, 80–90 °C (180–200 °F).

Etching:

All macroetchants are suitable for microetching if etching time is shortened.

No.	Etchant		Conditions	Remarks
Ni m1	Dist. water or ethanol (96%) Hydrochloric acid (1.19) Iron (III) chloride (Concentration variable)	20–100 ml 2–25 ml 5–8 g	5–6 s. Swab.	Ni-Fe, Ni-Cu, and Ni-Ag alloys. Ni-base superalloys. Monel.
Ni m2	Dist. water Nitric acid Glacial acetic acid (Concentration variable)	0 (10) ml 50 (38) ml 50 (100) ml	5–30 s. Use fresh only! !!! See Appendix A.	Grain boundaries. Pure types of Ni and alloys with high Ni content. Ni-Ti and Ni-Cu alloys.
Ni m3	Nitric acid (1.40) Hydrochloric acid (1.19)	20 ml 100 ml	5–6 s. Does not keep! !!! See Appendix A.	Ni-base superalloys.
Ni m4	a. Dist. water Potassium cyanide b. Dist. water Ammonium persulfate Hydrogen peroxide (3%)	95 ml 5 g 100 ml 10 g few drops	Secs to mins. Use fresh only! Mix equal amounts of (a) and (b). !!! See Appendix A.	Ni-Zn-Ag, Ni-Ag-Ni-Cu, and Ni-Al-Mo alloys. Nonmetallic inclusions are not attacked.

Ni m5	Hydrochloric acid (1.19) Chromium (VI) oxide	100 ml 0.01–1 g	Secs to mins. !!! See Appendix A.	Ni-Al, Mo-Ni, Ni-Ti alloys.
Ni m6	Nitric acid (1.40) Hydrofluoric acid (40%)	80 ml 3 ml	Secs to mins. Microsection should be heated in boiling water prior to etching. !!! See Appendix A.	Ni and Ni-base alloys. Ni-Cr alloys.
Ni m7	Methanol (95%) Nitric acid (1.40)	10 ml 10 ml	5 min.	WC-Mo_2C-TiC-Ni cemented carbides.
Ni m8	Ammonia water Hydrogen peroxide (30%)	85 ml 15 ml	5–15 s. Does not keep!	Ni-Zn alloys.
Ni m9	Dist. water Hydrochloric acid (1.19) Chromium (VI) oxide	50 ml 150 ml 25 g	5–20 s. !!! See Appendix A.	Ni-base superalloys. Especially suitable for Hastelloy types.
Ni m10	Hydrochloric acid (1.19) Nitric acid (1.40) Glycerol Glacial acetic acid (Concentration variable)	40 ml 30 ml 10 ml 20 ml	Secs to mins. Do not store! !!! See Appendix A. Without acetic acid	Ni-Fe and Ni-Al alloys. Ni-base superalloys.
Ni m11	Ethanol or methanol (95%) Hydrochloric acid (1.19) Copper (II) chloride	40–80 ml 40 ml 2 g	Secs to mins.	Ni-Cu alloys. Ni-base superalloys.
Ni m12	Dist. water Nitric acid (1.40) Hydrofluoric acid (40%)	80 ml 10 ml 10 ml	Secs to mins. !!! See Appendix A.	Ni-silicides.
Ni m13	Dist. water Nitric acid (1.40)	50 ml 50 ml	30–60 s. 90–100 °C (200–210 °F).	80Ni-20Cr and 35Ni-20Cr-45Fe alloys.
Ni m14	*Electrolytic:* Dist. water Potassium cyanide	100 ml 10 g	Approx 3 min, 6 V dc, Pt cathode. !!! See Appendix A.	Carbides in Ni-Cr alloys.
Ni m15	*Electrolytic:* Dist. water Sulfuric acid (1.84)	100 ml 2–50 ml	5–15 s, 6V dc, Pt cathode.	Ni and Ni-base alloys. Ni-Cu and Ni-Cr alloys. Carbide inclusions.

Ni m16	*Electrolytic:* Dist. water Ammonium persul- fate or chromium (VI) oxide	 100 ml 10 g 6 g	 Secs to mins, 6 V dc, Ni cathode. !!! See Appendix A.	 Ni and Ni-base alloys. Ni-Cr and Ni-Fe alloys. Cast alloys.
Ni m17	*Electrolytic:* Dist. water Nitric acid (1.40) Glacial acetic acid	 85 ml 10 ml 5 ml	 20–60 s, 1.5 V dc, Pt cathode. Do not store! !!! See Appendix A.	 Grain contrast in Ni. Ni-Ag, Ni-Al, Ni-Cr, Ni-Cu, Ni-Fe, and Ni-Ti alloys.
Ni m18	*Electrolytic:* Dist water Glycerol Hydrofluoric acid	 85 ml 10 ml 5 ml	 2–10 s, 2–3 V dc, Ni cathode. !!! See Appendix A.	 Ni-Al alloys.
Ni m19	*Electrolytic:* Dist. water Oxalic acid	 100 ml 10 g	 10–15 s, 6 V dc, stainless steel cath- ode.	 Ni-Au, Ni-Mo, and Ni-Cr alloys. Micro-inhomogene- ities in superalloys.
Ni m20	*Electrolytic:* Dist water Phosphoric acid (171) Possibly: Sulfuric acid (1.84)	 30 ml 70 ml 15 ml (max)	 5–60 s, 2–10 V dc, Ni cathode.	 Ni and Ni-base alloys. Ni-Cr, Ni-Fe alloys. Superalloys of the Nimonic type.
Ni m21	*Electrolytic:* Phosphoric acid (1.71) Sulfuric acid (1.84) Chromium (VI) oxide	 85 ml 5 ml 8 g	 5–30 s, 10 V dc, Pt cathode. !!! See Appendix A.	 Ni-base superalloys. Gamma precipitates. Ti and Nb micro- segregations.

Lead

Pb

Because of the low hardness and low recrystallization temperature of lead, deformation layers are difficult to avoid, and erroneous microstructures form easily. Preparation methods which do not generate mechanical strain or heat are preferred. Respiratory poison and stomach poisoning may occur upon ingestion of lead into the human body (!!! See Appendix A).

Macroetching

Preparation:

In general, either dressing with a special file for light and soft metals, or coarse grinding, is sufficient. Areas of the surface not of interest are covered with plastic spray.

Etching:

No.	Etchant		Conditions	Remarks
Pb M1	Dist. water Nitric acid (1.40) (Concentration variable)	80 ml 20 ml	10 min. Swab.	Grain contrast. Welded seams. Layers.
Pb M2	Glycerol Glacial acetic acid Nitric acid (1.40)	68 ml 16 ml 16 ml	Few min. Use fresh only! Useful as microetchant (few secs).	Grain contrast. Pb containing Sb.
Pb M3	a. Dist. water Ammonium molybdate b. Dist. water Nitric acid (1.40)	100 ml 15 g 42 ml 58 ml	Mix (a) and (b) in equal proportions. 10–30 s. Swab. Rinse in running water.	Removal of deformed layers. Pure and low-alloy Pb.

Microetching

Preparation:

Observe specifications for macroetching.

Grinding:

a. Dry on silicon carbide papers down to 600 grit; coat papers with wax.

b. Dressing by lathe, less than 25 μm per cut.

c. Prepare by microtome using cemented carbide blade or, even better, diamond blade; eliminates polishing.

Polishing:

a. Electrolytic: 300 ml perchloric acid (60%), 700 ml glacial acetic acid; mix drop by drop, cool with ice. 3–4 min at 15 ° C (60 °F). 8 V, dc. (!!! See Appendix A.)

b. Alumina slurry. Add 1 g ammonium acetate per litre in case of surface-film formation. Soft polishing cloth. Low rotation speed (200–300 rpm).

c. Hard lead, lead-rich white metals and type metals, specimen preparation same as for steel.

After polishing, rinse in distilled water. Tap water may etch surface.

Etching:

No.	Etchant		Conditions	Remarks
Pb m1	Glycerol or ethanol (96%) Glacial acetic acid Nitric acid (1.40) (Concentration variable)	84 (76) ml 8 (16) ml 8 ml	Few secs. Possibly 80 °C (180 °F). Use fresh only!	Pb, Pb-Sb, Pb-Cd, Pb-Ca alloys. Pb-containing Cu and bronzes. High-Pb white metals. Grain boundaries and precipitates.
Pb m2	Ethanol or methanol (95%) Nitric acid (1.40) Can also be used as macroetchant.	100 ml 1–5 ml	1–10 min. After etching, store in tap water for short time.	Pb and Pb alloys. Hard lead and high-Pb white metals and type metals. Pb solid-solution crystals are etched dark.

Pb m3	Dist. water or ethanol (96%) Hydrochloric acid (1.19) Iron (III) chloride (Concentration variable)	90 (90) ml 20 (30) ml 0 (10) g	1–10 min	Lead solder. Pb-Sb alloys and high-Pb white metals and type metals.
Pb m4	Dist. water Glacial acetic acid Nitric acid (1.40)	80 ml 15 ml 20 ml	14–30 min, 40 °C (100 °F). Use fresh only!	Lead solders. Pb-Sn alloys.
Pb m5	Glacial acetic acid Hydrogen peroxide (30%) (Concentration variable)	75 ml 25 ml	5–15 s.	Pure Pb. Pb-Sb, Pb-Ca alloys. Grain boundaries.
Pb m6	Dist. water Silver nitrate	100 ml 5–10 g	Secs to mins. Etch-rinsing.	All Pb alloys. Hard lead. High-Pb white metals and type metals. Friction metals. Bearing metals.
Pb m7	*Electrolytic:* Perchloric acid (60%) Glacial acetic acid Mix drop by drop in ice bath for cooling.	30 ml 70 ml	1 min, 1–2 V dc, Pb or Cu cathode. !!! See Appendix A.	Grain contrast.
Pb m8	*Electrolytic:* Dist. water Perchloric acid (60%)	40 ml 60 ml	10 s, 2 V dc, Pt anode, sample as cathode. !!! See Appendix A.	Pb. Pb-Sb and Pb-Sn alloys.

Plutonium, Thorium, Uranium

Pu
Th
U

Care is advised! Radioactive materials! (!!! See Appendix A.) Work should be carried out in glove boxes with proper ventilation or "hot cells" if critical radiation limit is surpassed. In any case, check radiation! These materials must not be inhaled and are to be handled by properly trained authorized personnel.

Macroetching

Preparation:

Coarse grinding down to 400 grit, wet on silicon carbide papers.

Etching:

No.	Etchant		Conditions	Remarks
Pu M1	Nitric acid (1.40)		1–5 s.	U and U alloys.
Pu M2	*Electrolytic:* Dist. water Phosphoric acid (1.71)	90 ml 10 ml	Up to 15 min, 20 V dc, Al cathode.	U and most U alloys.

Microetching

Preparation:

Grinding:

Wet on silicon carbide papers down to 600 grit.

Polishing:

a. Alumina slurry or diamond paste down to finest particle size.

b. Electrolytic:

1. Glacial acetic acid Perchloric acid (60%)	900 ml 100 ml	30 s to mins, 12–20 V dc, stainless steel cathode. !!! See Appendix A.	Especially for Pu-Al alloys.
2. Ethanol (96%) Phosphoric acid (1.71) Glycerol	300 ml 300 ml 300 ml	4–6 min, 20–30 V dc, Al cathode. !!! See Appendix A.	
3. Ethanol (96%) Ethylene glycol (1.11) Phosphoric acid (1.71)	445 ml 275 ml 275 ml	1–5 min, 18–20 V dc, stainless steel cathode. !!! See Appendix A.	

Etching:

Uranium is suitable for observation in polarized light.
Heat tinting at 100–200 °C (210–390 °F) for several hours gives good results for most types of uranium and uranium alloys.

No.	Etchant		Conditions	Remarks
Pu m1	Nitric acid (1.40) Glacial acetic acid Glycerol	30 ml 30 ml 30 ml	5–30 s !!! See Appendix A.	U and most U alloys.
Pu m2	Dist. water Nitric acid (1.40) or Hydrofluoric acid (40%) (Concentration variable)	100 (50) ml 38 (25) ml 1 (25) ml	Secs to mins. !!! See Appendix A.	U-Al alloys. UAl_2 (light blue). UAl_3 (yellow). UAl_4 (gray).
Pu m3	Phosphoric acid (1.71) Sulfuric acid (1.84) Nitric acid (1.40)	70 ml 25 ml 5 ml	Secs to mins.	U alloys
Pu m4	Hydrofluoric acid (40%) Nitric acid (1.40) Lactic acid (90%) or Dist. water	1 ml 30 ml 30 ml	5–30 s. !!! See Appendix A.	U-Be alloys. U-beryllides, U-Zr, and U-Nb alloys.
Pu m5	Glycerol Nitric acid (1.40) Hydrofluoric acid (40%)	40 ml 40 ml 10 ml	5–10 s. !!! See Appendix A.	Gamma phase stabilized U-Mo and U-Zr alloys.
Pu m6	Hydrofluoric acid (40%) Nitric acid (1.40)	50 ml 50 ml	Few sec. !!! See Appendix A.	Th and Th-base alloys.
Pu m7	Lactic acid (90%) Nitric acid (1.40) Hydrofluoric acid (40%)	30 ml 30 ml 3 drops	Secs to mins.	U-silicides.

No.	Etchant		Conditions	Remarks
Pu m8	*Electrolytic:* Glacial acetic acid Perchloric acid	90 ml 10 ml	5–15 min, 18–20 V dc, stainless steel cathode. !!! See Appendix A.	Pure U and Th.
Pu m9	*Electrolytic:* Methanol (95%) Ethylene glycol (1.11) Nitric acid (1.40)	20 ml 50 ml 5 ml	2 min, 0.05 A/cm^2, stainless steel cathode.	Pu and Pu-base alloys.

Lanthanum and Rare Earths (Lanthanides)

RE | Ce | Dy | Er | Eu | Gd | Ho | La | Lu | Nd | Pm | Pr | Sm | Tb | Tm | Yb

Cerium, dysprosium, erbium, europium, gadolinium, holmium, lanthanum, lutetium, neodymium, promethium, praseodymium, samarium, terbium, thulium, ytterbium.

Microetching

Preparation:

Grinding:

Wet on silicon carbide papers down to 600 grit. Exceptions: europium, lanthanum, and rare-earth–cobalt alloys grind dry.

Polishing:

With diamond paste and/or alumina slurry down to finest particle size on felt or velvet.

Etching:

No.	Etchant		Conditions	Remarks
RE m1	Glacial acetic acid Hydrogen peroxide (30%)	75 ml 25 ml	5–15 s.	For most RE metals and their alloys.
RE m2	Heat tinting in air, room temperature to 200 °C (390 °F)		Mins to hours	For most alloys of RE metals.
RE m3	Nitric acid (1.40) Glacial acetic acid Phosphoric acid (1.71) Lactic acid (90%) Sulfuric acid (1.84) (Concentration variable)	15 ml 10 ml 5 ml 20 ml 1 ml	10–15 s.	Dy, Er, Gd, Ho-base alloys. RE-Co alloys.
RE m4	Ethanol (96%) Nitric acid (1.40)	49 ml 1 ml	2–3 min. Possible pre-etching with solution made of 30 ml nitric acid (1.40) and 70 ml glycerol. Heat solution in water bath.	Pure Gd. RE-Co alloys. Grain-boundary etch.

Sn

Tin

Because of the tendency of tin toward deformation and low recrystallization temperature, deformation layers may also form in pure tin as well as in some of its alloys. Artifacts and mechanical twinning occur commonly. Use preparation methods which do not generate mechanical strain and heat.

Macroetching

Preparation:

Coarse grinding (wet) on silicon carbide paper is often sufficient. A special file can also be used for light and soft metals. Possibly diamond polishing down to 6 μm.

Etching:

No.	Etchant		Conditions	Remarks
Sn M1	Sat. aqueous sol. of ammonium polysulfide		20–30 min. Wipe off with cotton after etching.	For all types of Sn and Sn-base alloys. Grain distribution.
Sn M2	Dist. water Hydrochloric acid (1.19) Iron (III) chloride	100 ml 2 ml 10 g	30 s to 5 min	Sn-rich bearing metals and white metals.

Microetching

Preparation:

Observe instructions for macroetching.

Grinding:

a. Wet on silicon carbide papers down to 600 grit (coat with wax).

b. Prepare by microtome using cemented carbide blade or diamond blade; eliminates polishing.

c. For high-tin white metals, use same method as for steel.

Polishing:

a. Electrolytic:

1. Dist. water Perchloric acid (70%) Glacial acetic acid	13 ml 63 ml 300 ml	Approx 10 min, 20–30 V dc, Sn cathode. !!! See Appendix A.
2. Dist. water Fluoboric acid (35%) Sulfuric acid (1.84)	780 ml 200 ml 20 ml	3–5 s, 20–40 °C (70–100 °F), 15–17 V dc, stainless steel cathode. !!! See Appendix A.

b. With alumina slurry (possible etch-polishing sequence) on Microcloth, or with diamond paste (particle size 6–0.25 μm). Subsequently on soft cloth with a suspension of 120 ml hot distilled water, 20 ml aqueous solution of tartaric acid (5%), and 1 g magnesia. Suspension should be filtered before use (through a fine nylon cloth or cotton material).

Etching:

No.	Etchant		Conditions	Remarks
Sn m1	Ethanol or methanol (95%) or dist. water Hydrochloric acid (1.19)	100 ml 2–5 ml	Few mins.	Pure Sn, Sn-Cd, Sn-Fe, Sn-Pb, and Sn-Sb-Cu alloys.
Sn m2	Dist. water or methanol (5%) Hydrochloric acid (1.19) Iron (III) chloride	100 ml 5–25 ml 10 g	10–30 s (possibly up to 5 min).	Sn-rich bearing metals and white metals. Sn-Cu alloys and Bi-Sn eutectic.
Sn m3	Glycerol Hydrofluoric acid (40%) Nitric acid (1.40)	25 ml 2 ml 1 drop	1 min. !!! See Appendix A.	Sn layers on steel. Sn-Pb alloys.
Sn m4	Ethanol (96%) Nitric acid (1.40) (Nital)	100 ml 1–5 ml	2–3 s. Possibly followed by stronger nital etchant.	Pure Sn. Sn-rich alloys with Cd, Fe, Sb, Cu.
Sn m5	Dist. water Ammonium persulfate	100 ml 5–10 g	Secs to mins.	Sn layers. Sn bearing metals.
Sn m6	Cold sat. aqueous sol. of sodium thiosulfate Potassium metabisulfite	50 ml 5 g	60–90 s.	Grain contrast (color) in pure Sn and most Sn alloys.
Sn m7	Ethanol (96%) Picric acid (Picral)	100 ml 4 g	Secs to mins.	Sn layers on steel and cast iron.
Sn m8	*Electrolytic:* Dist. water Sulfuric acid (1.84)	 80 ml 20 ml	Secs to mins, 30 V dc, Al cathode.	Pure Sn. Withdraw sample while current is on.
Sn m9	*Electrolytic:* Dist. water Glacial acetic acid Perchloric acid (70%)	 25 ml 300 ml 50 ml	10 min, 20–30 V dc, Sn cathode. !!! See Appendix A.	Sn and Sn-rich alloys.

Titanium

Ti

Pure titanium is soft and deformation layers easily form. Titanium alloys usually do not pose any particular problems.

Macroetching

Preparation:

Wet grinding on silicon carbide papers down to 600 grit.

Etching:

No.	Etchant		Conditions	Remarks
Ti M1	Dist. water Nitric acid (1.40) Hydrofluoric acid (40%) Content of hydrofluoric acid may be decreased	50 ml 40 ml 10 ml	5–8 min, 60–80 °C (140–180 °F). !!! See Appendix A.	Ti and Ti-base alloys. Ti-Al-Mo alloys.
Ti M2	Dist. water Hydrofluoric acid (40%) Hydrogen peroxide (30%)	30 ml 10 ml 60 ml	Swab until desired contrast is obtained.	Iodide Ti.
Ti M3	Dist. water Hydrofluoric acid (40%) Iron (III) nitrate Oxalic acid	200 ml 2 ml 10 g 35 g	Secs to mins, 50–60 °C (120–140 °F). !!! See Appendix A.	Welded seams.
Ti M4	Dist. water Hydrochloric acid (1.19)	50 ml 50 ml	Secs to mins.	Differentiation of alpha and beta Ti.

Microetching

Preparation:

Grinding:

a. Same as for macroetching. Starting with 400 grit, coat papers with wax.

b. Prepare by microtome using cemented carbide blade. Only final polishing needed after this treatment. Diamond blades are not suitable, because they tend to leave a rough surface.

Polishing:

a. Electrolytic:

1. Glacial acetic acid Perchloric acid (70%)	900 ml 60 ml	1–5 min, 20–60 V dc, stainless steel cathode. !!! See Appendix A.
2. Ethanol (96%) *n*-Butyl alcohol Aluminum chloride Zinc chloride	90 ml 10 ml 6 g 28 g	1–6 min, 20–25 V dc, stainless steel cathode. !!! See Appendix A.

b. With diamond paste or alumina slurry down to finest particle size. Possible multiple polishing with one of the electrolytes mentioned above or etch-polishing sequence with one of the electrolytes mentioned above or etch-polishing sequence with one of the microetchants listed.

c. Chemical:

1. With macroetchant Ti M2		
2. With microetchant Ti m1		Approx 5–20 s.
3. Hydrofluoric acid (40%) Nitric acid (1.40) Lactic acid (90%)	10 ml 10 ml 30 ml	Secs to mins. !!! See Appendix A.

Etching:

Titanium is suitable for observations in polarized light.

No.	Etchant		Conditions	Remarks
Ti m1	Dist. water Hydrofluoric acid (40%) Nitric acid (1.40) (Kroll's reagent)	100 ml 1–3 ml 2–6 ml	3–10 s. !!! See Appendix A.	For many Ti raw materials. Especially for Ti-Al-V alloys.
Ti m2	Dist. water Hydrofluoric acid (40%) Nitric acid (1.40) (Concentration variable)	85 (96) ml 10 (2) ml 5 (2) ml	2–3 s. !!! See Appendix A.	Most types of Ti and Ti alloys. Ti-Mn, Ti-V-Cr-Al alloys.
Ti m3	Dist. water Hydrofluoric acid (40%) or Sulfuric acid (1.84)	100 ml 1–10 ml 1–10 ml	Secs to mins. Glycerol instead of water decreases attack rate. !!! See Appendix A.	Ti alloys. Alpha Ti is attacked.
Ti m4	Dist. water Hydrogen peroxide (30%) Aqueous sol. of potassium hydroxide (Concentration variable)	78 (20) ml 15 (5) ml 12 (10) ml	3–20 s. Swab.	Ti. Ferro-Ti. Ti-Al-V-Sn alloys. Alpha Ti is attacked; beta Ti is left unchanged.
Ti m5	Dist. water Glycerol Nitric acid (1.40) Hydrofluoric acid (40%)	20 ml 45 ml 25 ml 1 ml	3–20 s. Swab. !!! See Appendix A. Use 3 ml hydrochloric acid (1.19) instead of water.	Ti-Al-Sn and Ti-Al-Ni alloys. Ti-Si alloy.
Ti m6	Hydrofluoric acid (40%) Nitric acid (1.40) Lactic acid (90%) (Concentration variable)	1 ml 30 ml 30 ml	5–30 s. Do not store! See Appendix A.	Attacks hydrides in Ti preferentially.
Ti m7	Dist. water Hydrofluoric acid (40%) Hydrogen peroxide (30%)	100 ml 2 ml 5 ml	30–60 s. !!! See Appendix A.	Ti and Ti alloys. Grain boundary etchant.
Ti m8	*Electrolytic:* Dist. water Methanol (95%) Ethylene glycol (1.11) Perchloric acid (70%)	25 ml 390 ml 350 ml 35 ml	10–40 s, 5–10 °C (40–50 °F), 30–50 V dc, stainless steel cathode. !!! See Appendix A.	Pure Ti.

Ti m9	*Electrolytic:* Glacial acetic acid Perchloric acid (70%)	80 ml 5 ml	1–5 min, 20–60 V dc, stainless steel cathode. !!! See Appendix A.	Pure Ti and Ti-base alloys.
Ti m10	*Electrolytic:* Dist. water Ethanol (96%) Lactic acid (90%) Phosphoric acid (1.71) Citric acid Oxalic acid	35 ml 60 ml 10 ml 5 ml 5 g 5 g	10 s, 30–50 V dc, stainless steel cathode. Possible dilution with equal amount of glycerol. Then, use 130 V dc, 60 s. !!! See Appendix A.	Ti and Ti-base alloys. Tint etching. Precipitates.

If a deposit forms during etching, immediately clean sample in aqueous solution of sulfuric acid (10–30%).

Zn

Zinc

During metallographic preparation of zinc, deformation layers may easily form. False microstructures are also common. Therefore, mechanical strain and heat formation should be avoided during preparation.

Macroetching

Preparation:

Coarse grinding is sufficient. Possibly pre-polishing with diamond paste, particle size 6 μm.

Etching:

No.	Etchant		Conditions	Remarks
Zn M1	Dist. water Hydrochloric acid (1.19) or concentrated hydrochloric acid only or concentrated nitric acid (1.40) only	50 ml 50 ml	Approx 15 s. Rinse off film under running water.	Pure Zn. Zn alloys without Cu. Cast materials.
Zn M2	Dist. water Chromium (VI) oxide Sodium sulfate (anhydrous) (Palmerton's reagent)	100 ml 20 g 1.5 g	Secs to mins. If using sodium sulfate containing crystal water ($Na_2SO_4 \cdot 10H_2O$) increase to 3.5 g. !!! See Appendix A.	Zn alloys containing Cu. Dilute with 100 ml water if used as microetchant.

Microetching

Preparation:

See recommendations for macroetching.

Grinding:

a. Wet on silicon carbide papers down to 600 grit (coat papers with wax).

b. Prepare by microtome using cemented carbide blade or diamond blade; eliminates polishing.

Polishing:

a. With alumina slurry on Microcloth, or with diamond paste (particle size: 6, 3, 1, and possibly 0.25 μm, followed by polishing with a suspension of 120 ml warm distilled water, 20 ml of aqueous solution of ammonium tartrate (5%), and 1 g magnesium oxide. Before use, suspension should be filtered through a fine nylon or cotton cloth.

b. To eliminate the deformed layer originated in grinding, a pre-polishing may be useful. Soap is added to an alumina slurry, and the specimen is polished with this mixture on a soft wool cloth. Attack-polishing may also be suitable. Here, mechanical polishing with alumina or diamond alternates with etching, employing etchant Zn m1. The following sequence has proven successful: 4 min diamond (6 μm), 3.5 min etch, 6 min diamond (3 μm), 1.5 min etch, 8 min diamond (1 μm), 30 s etch, 10 min diamond (0.25 μm), 10 s etch, polishing with alumina slurry No. 3 (0.05 μm, Linde B), 5 sec etch.

c. Electrolytic:

1. Dist. water Chromium (VI) oxide	100 ml 20 g	40–50 s, 60 V dc, Pt or Zn cathode. !!! See Appendix A.
2. Phosphoric acid (1.71) Ethanol (96%)	50 ml 50 ml	1 h and longer, 4–6 V dc, stainless steel cathode. !!! See Appendix A.

d. Chemical:

Dist. water Nitric acid (1.40) Sodium sulfate Chromium (VI) oxide	100 ml 5 ml 1.5 g 20 g	Up to 30 min. Layer of reaction products is water soluble. !!! See Appendix A.

Etching:

Zinc is suitable for observation in polarized light.

No.	Etchant		Conditions	Remarks
Zn m1	Dist. water Chromium (VI) oxide Sodium sulfate	100 ml 20 g 1.5g	2–3 min. For alloys containing Cu, use half the amount of sodium sulfate. Rinse in: 100 ml dist. water 20 g chromium (VI) oxide !!! See Appendix A.	For most types of Zn and Zn alloys. Especially for rolled Zn containing Pb and Zn-Cu alloys. If chromium (VI) oxide content is lowered to 5 g and the sodium sulfate to 0.5 g, etchant can be used for pressure-cast parts, and Zn platings.
Zn m2	Dist. water, ethanol, methanol, or amyl alcohol (95%) Nitric acid (1.40)	100 ml 0.5–1 ml	Secs to mins. Etching solution should not be older than 1 h. To avoid stains, rinse specimen in aqueous sol. of chromic acid (20%) after etching	Zn-Fe layers of galvanized steel or ferrous materials. Zn-Al and Zn-Cr alloys. Grain boundaries.

Zn m3	Dist. water Sodium hydroxide	100 ml 10 g	1–5 s.	Technically pure Zn. Zn-Co and Zn-Cu alloys. Generally for low-alloy Zn.
Zn m4	Aqueous sol. of sodium thiosulfate Potassium metabisulfite	50 ml 1 g	30 s.	Tint etching of Zn and low-alloy Zn.
Zn m5	Dist. water or ethanol (96%) Hydrochloric acid (1.19) Possibly increase water content	100 ml 1–5 ml	Secs to mins.	Zn and Zn-rich alloys. Zn-Cu-Al alloys.
Zn m6	Dist. water Ethanol (96%) Picric acid (Concentration variable)	70 ml 30 ml 0.3 g	Secs to mins.	Zn alloys with metals of the iron group. Zn platings.
Zn m7	Dist. water Hydrochloric acid (1.19) Iron (III) chloride	1250 ml 30 ml 4 g	Secs to mins.	Zn alloys with much nobler constituents, for example, Cu, Ag, Au.
Zn m8	Dist. water Saturated sol. of copper (II) nitrate Potassium hydroxide Potassium cyanide Citric acid (Concentration variable) (Schramm's reagent)	900 ml 60 ml 122 g 75 g 6.5 g	Secs to mins. Let precipitating crystals settle. !!! See Appendix A.	Fe, Pb, Mg, Cu, and Ni-Zn alloys. Zn-rich phase dark. Also suited as macroetchant.
Zn m9	Dist. water Citric acid Ammonium persulfate	100 ml 1 g 11 g	Secs to mins.	Fiber structure in extruded Zn alloys.
Zn m10	*Electrolytic:* Dist. water Sodium hydroxide	100 ml 25 g	15 min, 6 V dc, Cu cathode.	Pure and low-alloy Zn.
Zn m11	*Electrolytic:* Dist. water Chromium (VI) oxide	100 ml 10 g	Secs to mins, 12 V dc, Pt cathode. !!! See Appendix A.	Zn-Cu alloys to distinguish between gamma and epsilon phase.

Chapter 3:
Preparation of Special Ceramics and Cermets (Ceramography)

Most engineering ceramics and cermets are examined by reflected-light microscopy. Silicate ceramics, however, are more frequently examined as thin sections, using transmitted-light microscopy. Preparing suitable specimens requires certain deviations from the usual metallographic procedure (refer to Chapters 1 and 2), thereby creating the expression "ceramography."

Porosity and plucking are the greatest problems in preparing ceramics. The hardness differential between the ceramic and metallic components is the major problem when preparing cermets. Macroscopic examinations of special ceramics and cermets is normally not neccessary. In any case, the preparation methods for macroscopic investigations are identical to those for microscopic observations.

Ceramics are often porous and contain cracks and therefore should be impregnated with a resin to preserve the specimen (refer to Chapter 1). Toxic materials such as beryllium oxide, uranium compounds, and plutonium compounds should be handled carefully according to precautions in Appendix A. The need for these precautions is indicated by the (!!!) notation.

Preparation:

Grinding:

a. Wet on silicon carbide papers down to 320 grit or 600 grit.
b. Beginning with grit size 320, fine grinding on diamond wheels.
c. Lapping with silicon carbide slurries of particle size 7–1 μm on a cast iron wheel.

Polishing:

a. With diamond paste of particle size 15–1 μm for one to two days or longer on automatic polishers (for example, vibrating polishers).
b. With diamond paste of particle size 6–1 μm on a fast-rotating hardwood wheel.
c. With diamond paste of particle size 6–1 μm with a stick of rosewood of approx 15 mm diameter in a hand drill machine. Fast polishing action. Polished surface area is equal to diameter of wood stick and is not flat. Wood stick should not touch embedding material (smearing).
d. With alumina slurry No. 1 (5 μm) and No. 1C (1 μm, Linde C) on a fast-rotating wheel covered with strong, smooth felt.
e. Alumina slurry (0.05 μm; 30 g in 140 ml hydrogen peroxide, 3%) on a fast-rotating

wheel covered by moistened velvet or felt. Finish with high pressure. Clean sample under running water immediately after polishing.

f. With electrically conductive ceramics (for example, carbides), electrolytic polishing is possible. The etchants listed below are often suitable for electrolytic polishing. Polishing times are six to ten times the etching time.

Oxides

No.	Etchant		Conditions	Remarks
O m1	Hot etching in air		2 h, 1100–1500 °C (2000–2730 °F), 4 min.	Al_2O_3. Al_2O_3-MgO mixture. SnO_2.
O m2	Hot etching in pure, dry hydrogen		1–3 min, 700–1600 °C) (1300–2900 °F) 500 torr, 3 min, 1200 °C (2200 °F). !!! See Appendix A.	Al_2O_3. BeO (!!!). UO_2 (!!!).
O m3	Hot etching in dry hydrogen		500 torr, 3–10 min, 1200 °C (2200 °F).	UO_2 (!!!).
O m4	Hot etching in a 1:1 mixture of water vapor and argon		2 h, 1250 °C (2300 °F).	UO_2 (!!!). Grain size.
O m5	Thermal etching		1 h, 10 torr, 1600 °C (2900 °F).	UO_2 (!!!).
O m6	Potassium hydrogen fluoride melt		5–10 min, molten salt in Pt crucible. !!! See Appendix A.	Al_2O_3. Al_2SiO_5. Rinse sample in phosphoric acid.
O m7	Potassium hydrogen sulfate melt		3–10 min, molten salt in Pt crucible. 15–25 s. !!! See Appendix A.	Cr_2O_3. CeO_2. Al_2O_3.
O m8	Hydrochloric acid (1.19)		3 s to 6 min	CaO, MgO.
O m9	Dist. water Hydrochloric acid (1.19)	10 ml 10 ml	Secs to mins. Boiling.	ThO_2-Y_2O_3 mixtures (!!!)
O m10	Hydrofluoric acid (40%)		2 s to 6 min. !!! See Appendix A.	BeO (!!!). Zr_2O_3. BaO, MgO. $Ca_xZr_{1-x}O_y$.
O m11	Dist. water Hydrofluoric acid 40% (Concentration variable)	10 (100) ml 110 (1) ml	10–20 min, 60–80 °C (140–180 °F) !!! See Appendix A.	Al_2O_3, SiO_2. BeO-UO_2-Y_2O_3 mixtures (!!!).
O m12	Dist. water Hydrochloric acid (1.19) Sat. aqueous sol. of copper (II) sulfate	10 ml 10 ml 10 ml	5 s.	Eu_2O_3.

O m13	Phosphoric acid (1.71)		1–30 min. Boiling.	MgO. ThO_2 (!!!). Al_2NiO_4, PuO_2 (gamma, sintered) (!!!). Y_2O_3-ZrO_2 and Sm_2O_3-ZrO_2 mixtures.
O m14	Dist. water Phosphoric acid (1.71)	15 ml 85 ml	5 min to 2 h. Boiling.	Al_2O_3. Relief formation.
O m15	Dist. water Nitric acid (1.40) (Concentration variable)	10 (100) ml 10 (15) ml	3 s to 5 min, room temperature to 60 °C (140 °F).	MgO, UO_2 (!!!).
O m16	Sulfuric acid (1.84)		1 min to 2 h. 1–2 min, 60 °C (140 °F).	Al_2O_3, UO_2 (!!!). ThO_2 (!!!), Al_2O_3-MgO mixtures.
O m17	Dist. water Sulfuric acid (1.84)	50 (10) ml 50 (1) ml	1–5 min, boiling. 1–5 min, 30 °C (85 °F). 10 sec, 30 °C (85 °F).	ZrO_2. U_3O_8 (!!!). Concentration in parentheses for Nd_2O_3.
O m18	Dist. water Glacial acetic acid (1.84)	100 ml 5 ml	Secs to mins.	ZnO.
O m19	Hydrofluoric acid (40%) Nitric acid (1.40) (Concentration variable)	10 (1) ml 50 (30) ml	5–10 min, 60–80 °C (140–180 °F). !!! See Appendix A. 10 min, boiling. 30 s to 1 min, boiling.	UO_2 (!!!). $Th_xU_yO_2$ (!!!). PuO_2, cast (!!!). PuO_2 (gamma, sintered) (!!!)
O m20	Dist. water Nitric acid (1.40) Hydrofluoric acid (40%)	20 (100) ml 20 (90) ml 10 (10) ml	5–15 min. !!! See Appendix A.	CeO_2, $SrTiO_3$. Al_2O_3. ZrO-ZrC mixtures.
O m21	Hydrochloric acid (1.19) Hydrofluoric acid (40%)	10 ml 3 ml	7 min to 2 h. !!! See Appendix A.	$BaTiO_3$. $BaTi_3O_7$.
O m22	Hydrogen peroxide (30%) Sulfuric acid (1.84) (Concentration variable)	1 (10) ml 10 (1) ml	1–11 min	UO_2. UO_2-PuP_2. UO_2-U_4O_9 and UO_2-CeO_2 mixtures (!!!).
O m23	Dist. water Nitric acid (1.40) Hydrogen peroxide (30%)	9 ml 1 ml 2 ml	Secs to mins.	U_4O_9 (!!!).
O m24	Sat. aqueous sol. of sodium sulfide		15–70 sec	CaO.

O m25	Dist. water Ammonium hydrogen fluoride	100 ml 25–50 g	3–5 min, 60 °C (140 °F). Prewarm sample in water.	BeO (!!!).
O m26	Lactic acid (90%) Nitric acid (1.40) Hydrofluoric acid (40%)	90 ml 15 ml 5 ml	10 min to 1 h, 65 °C (150 °F). !!! See Appendix A.	BeO (!!!).
O m27	Dist. water Nitric acid (1.40) Hydrofluoric acid (40%) Cerium (IV) nitrate	80 ml 20 ml 3 drops 1 g	Secs to mins. !!! See Appendix A.	UO_2-PuO_2 mixtures (!!!).
O m28	Methanol (95%) Hydrochloric acid (1.19) Hydrofluoric acid (40%)	100 ml 3 ml 0.5 ml	1 min. !!! See Appendix A.	MgO-Al_2O_3-SiO_2-ZrO_2 mixtures.
O m29	*Electrolytic:* Dist. water Sulfuric acid (1.84) Glacial acetic acid Chromium (VI) oxide	 15 ml 10 ml 20 ml 1 g	10–15 V dc, 1 A/cm^2, stainless steel cathode. 60–90 s (etch). 30–50 s (polish). !!! See Appendix A.	UO_2 (!!!).
O m30	*Electrolytic:* Dist. water Conc. aqueous oxalic acid Conc. aqueous citric acid Lactic acid (90%) Ethanol (96%) Phosphoric acid (1.71)	 35 ml 30 ml 30 ml 10 ml 60 ml 5 ml	3 s to 6 min, 17–20 V dc, stainless steel cathode. !!! See Appendix A.	Nb oxides. NbO (blue), NbO_2 (cyan), Nb_2O_5 (reddish-brown).
O m31	*Electrolytic:* Dist. water Hydrofluoric acid (40%) Glacial acetic acid	 60–70 ml 25 ml 25 ml	30–45 s, 2–4 mA/cm^2, 6–12 V dc, stainless steel cathode !!! See Appendix A.	NiO.

Carbides

No.	Etchant	Conditions	Remarks
C m1	Air	10 min to 24 h, 20–25 °C (70–80 °F).	ThC (!!!).
C m2	Dry, high-purity argon	20 min to 24 h, 20–25 °C (70–80 °F).	ThO_2 (!!!).

C m3	Hydrogen sulfide		12–30 s. !!! See Appendix A.	KC.
C m4	Thermal etching		≤ 10^{-3} torr, 1200 °C (2200 °F).	SiC.
C m5	Sodium or potassium bicarbonate melt		10 min, molten salt. !!! See Appendix A.	SiC.
C m6	Sodium tetraborate melt		Few mins. !!! See Appendix A.	SiC.
C m7	Nitric acid (1.40)		Secs to mins.	UC-$Cr_{23}C_6$ mixtures (!!!).
C m8	Dist. water Nitric acid (1.40)	10 ml 10 ml	Secs to mins. 1–45 min.	(Th,U)C (!!!) with U/Th ratio < 3. ThC_2 (!!!).
C m9	Nitric acid (1.40) Hydrofluoric acid (40%)	10 (30) ml 10 (10) ml	Secs to mins. !!! See Appendix A.	(Fe,Si)C. Concentration in parentheses for TaC.
C m10	Dist. water Nitric acid (1.40) Hydrofluoric acid (40%)	10 ml 10 ml 10 ml	Secs to mins. Glycerol instead of dist. water. !!! See Appendix A.	CrC, HfC. VC, (Al,Ti)C.
C m11	Hydrofluoric acid (40%) Nitric acid (1.40) Lactic acid (90%)	10 ml 10 ml 20 ml	Secs to mins. !!! See Appendix A.	Ta(C,N,O).
C m12	Dist. water Nitric acid (1.40) Glacial acetic acid	10 ml 10 ml 10 ml	Secs to mins. Swab. Few drops of oxalic acid or hydrofluoric acid (40%). !!! See Appendix A.	UC, U(C,O), UC-UC_2, UC_2-U_2C_3 mixtures. (U,Pu)C high in C. U(C,N), UC-ZrC mixtures (!!!).
C m13	Dist. water Nitric acid (1.40) Hydrochloric acid (1.19)	25 ml 25 ml 1 ml	Secs to mins.	ThC (!!!).
C m14	Nitric acid (1.40) Hydrochloric acid (1.19) Sulfuric acid (1.84)	10 ml 10 ml 10 ml	Secs to mins. Hydrofluoric acid (40%) can be used instead of hydrochloric acid. !!! See Appendix A.	TaC.
C m15	Hydrochloric acid (1.19) Hydrogen peroxide (30%)	10 ml 10 ml	Secs to mins.	WC.
C m16	Dist. water Formic acid (1.22)	10 ml 10 ml	4 s.	UC-Pu mixtures (!!!).

	Etchant	Amount	Conditions	Use
C m17	Glacial acetic acid Phosphoric acid (1.71)	10 ml 10 ml	8–10 min	$(Th,U)C_2$ (!!!).
C m18	Ethylene glycol (1.11) Ethanol (96%) Phosphoric acid (1.17)	8 ml 5 ml 5 ml	5–60 s. Swab. !!! See Appendix A.	(U,Pu)C (!!!).
C m19	Dist. water Sodium hydroxide Potassium ferricyanide	100 ml 10 ml 10 g	2–20 min, 50 °C (120 °F). !!! See Appendix A.	MoC_2, CrC_2.
C m20	*Electrolytic:* Aqueous sol. (8%) of sodium hydroxide Phosphoric acid (1.71) Sulfuric acid (1.84) Copper (II) sulfate	 80 ml 80 ml 10 ml 10 g	30 s, 25 °C (80 °F), 3.5 V dc, 0.9 A/cm^2, Cu cathode.	TiC.
C m21	*Electrolytic:* Same as C m12		Up to 50 s, 80 mA/cm^2, stainless steel cathode.	PuC (dendritic) (!!!).
C m22	*Electrolytic:* Same as C m17		1 min, 30–35 V, dc, stainless steel cathode.	ThC_2 (!!!).
C m23	*Electrolytic:* Sat. aqueous sol. of ammonium acetate		3–5 s, 5–7 V, dc, stainless steel cathode.	(U,Pu)C high in U. UC-PuN mixtures (!!!).
C m24	*Electrolytic:* Lactic acid (90%) Nitric acid (1.40) Hydrofluoric acid (40%)	 50 ml 30 ml 8 ml	Secs to mins. 20–25 °C (70–80 °F), 17–20 V dc, stainless steel cathode. !!! See Appendix A.	NbC, NbC_2.
C m25	*Electrolytic:* Dist. water Potassium hydroxide	 10 ml 2 g	2–30 s, 2 V dc, 30–60 mA/cm^2, Pt cathode. Move specimen.	TiC, TaC.
			20 s, 6 V dc, 1 A/cm^2.	SiC.
	Potassium hydroxide	0.1 g	40 V dc, 3 A/cm^2, stainless steel cathode.	B_4C.
C m26	*Electrolytic:* Dist. water Glacial acetic acid Chromium (VI) oxide	 7 ml 133 ml 25 g	1 min, 20 V dc, stainless steel cathode. !!! See Appendix A.	PuC (!!!).
C m27	*Electrolytic:* Dist. water Butyl glycol Phosphoric acid (1.71)	 1 ml 6 ml 3 ml	20–40 s, 5–10 V dc, 10–15 mA/cm^2, stainless steel cathode.	(U,Pu)C high in Pu (!!!).

Nitrides

No.	Etchant		Conditions	Remarks
N m1	Hot etching in dry high-purity nitrogen		5 h, 1600 °C (2900 °F). Variable.	Si_3N_4.
N m2	Hot etching in dry high-purity hydrogen		3 h, 1650 °C (3000 °F). !!! See Appendix A.	UN (!!!).
N m3	Thermal etching		18 h, 1650 °C (3000 °F) $\leq 10^{-5}$ torr	UN (!!!).
N m4	Potassium carbonate Sodium fluoride	95.4 g 12 g	1–4 min, molten salt. !!! See Appendix A.	Si_3N_4.
N m5	Phosphoric acid (1.71)		5–15 min. Boiling.	Si_3N_4. UN (!!!).
N m6	Hydrofluoric acid (40%)		10–15 min. !!! See Appendix A.	Si_3N_4.
N m7	Dist. water Glacial acetic acid Nitric acid (1.40)	10 ml 10 ml 10 ml	Secs to mins.	(Al,Ti)N.
N m8	Lactic acid (90%) Nitric acid (1.40)	10 ml 10 ml	30 s to 1 min. After etching for 30 s, add 7 drops of hydrofluoric acid (40%). !!! See Appendix A. 3 min, 40 °C (100 °F) without hydrofluoric acid.	UN, UN-U_2N_3 mixtures (!!!). UN-U(N_2O)-U_2N_3 mixtures (!!!)
N m9	Dist. water Glacial acetic acid Chromium (VI) oxide	60 ml 600 ml 50 g	30 s, 10 °C (50 °F). !!! See Appendix A.	UN (!!!).
N m10	*Electrolytic:* Sulfuric acid (1.84) Phosphoric acid (1.71) Glycerol	 10 ml 30 ml 30 ml	3–10 s, 4 V dc, stainless steel cathode. !!! See Appendix A.	UN (!!!).
N m11	*Electrolytic:* Glacial acetic acid Chromium (VI) oxide	 18 ml 1 g	4 s, 40 V dc, stainless steel cathode. !!! See Appendix A.	UN-U_2N_3 mixtures (!!!).
N m12	*Electrolytic:* Dist. water Ethanol (96%) Sat. aqueous sol. of oxalic acid Sat. aqueous sol. of citric acid Lactic acid (90%) Phosphoric acid (1.71)	 35 ml 60 ml 3 ml 3 ml 10 ml 5 ml	Secs to mins. 17–20 V dc, stainless steel cathode. !!! See Appendix A.	NbN (yellow). Nb_2N (light red).

Borides

No.	Etchant		Conditions	Remarks
B m1	Lactic acid Nitric acid (1.40) Hydrofluoric acid (40%)	30 ml 10 ml 10 ml	Secs to mins. Instead of lactic acid, 10 ml glycerol can be used.	ZrB_2. TiB_2.
B m2	Hydrochloric acid (1.19) Nitric acid (1.40)	10 ml 10 ml	1–5 min 40 °C (100 °F), vapor etching.	CrB_2. MoB_2.
B m3	Hydrochloric acid (1.19) Nitric acid (1.40)	100 ml 10 ml	15 s	TiB_2.
B m4	Hydrochloric acid (1.19) Nitric acid (1.40) Hydrofluoric acid (40%)	6 ml 2 ml 1 ml	9 s, 30–40 °C (85–100 °F). !!! See Appendix A.	HfB_2-NbB_2 mixtures.
B m5	Dist. water Hydrofluoric acid (40%) Nitric acid (1.40)	10 ml 10 ml 10 ml	Secs to mins. !!! See Appendix A.	ZrB_2.
B m6	Dist. water Sulfuric acid (1.84)	10 ml 1 ml	15 s.	TiB_2.
B m7	*Electrolytic:* Dist. water Sodium hydroxide	10 ml 1–2 g	Secs to mins, 10–15 V dc, stainless steel cathode.	TaB_2. LaB_4.

Phosphides and Sulfides

No.	Etchant		Conditions	Remarks
P m1	Lactic acid (90%)		Secs to mins.	PuP (!!!), PuS (!!!).
P m2	Dist. water Hydrofluoric acid (40%) Nitric acid (1.40) Sulfuric acid (1.84)	10 ml 10 ml 20 ml 20 ml	Secs to mins. !!! See Appendix A.	UO (!!!).
P m3	Hydrochloric acid (1.19)		6–12 s. Boiling, vapor etching.	CdS.
P m4	Hydrogen peroxide (30%) Sulfuric acid (1.84)	10 ml 1 ml	Secs to mins.	US (!!!).

P m5	Dist. water Hydrochloric acid (1.19) Dimethylene thiourea	30 ml 10 ml 1 g	1–10 min, 60 °C (140 °F)	PbS.

Cermets

No.	Etchant		Conditions	Remarks
Ct m1	Hot etching in dry, high-purity hydrogen		2.5 h, 1500 °C (2730 °F). !!! See Appendix A.	UO_2-Mo (!!!).
Ct m2	Hydrogen sulfide		15–30 s, room temperature. !!! See Appendix A.	UO_2-Cr (!!!).
Ct m3	a. Dist. water Potassium ferricyanide b. Dist. water Potassium or sodium hydroxide (Concentration variable)	100 ml 10 g 100 ml 10 g	Mix (a) and (b) in ratio 1:1 before use. 1–4 min. !!! See Appendix A.	ZrO_2-W. ThO_2-W (!!!). W_2C-W. UC-Cr (!!!). UC-Fe (!!!). UC-Ni (!!!). UC-UFe_2 (!!!).
Ct m4	Nitric acid (1.40)		Secs to mins.	$Cr_{23}C_6$-UFe_2 (!!!). US-U (!!!). UC_2-UNi_5 (!!!).
Ct m5	Dist. water Nitric acid (1.40) Hydrofluoric acid (40%)	50 (10) ml 30 (10) ml 10 (10) ml	Secs to mins. !!! See Appendix A.	TiN-Co, TiN-Fe. TiN-Mo, TiN-W. Concentrations in parentheses for HfC-Hf.
Ct m6	Dist. water Nitric acid (1.40) Glacial acetic acid	10 ml 10 ml 10 ml	Secs to mins. Possibly several drops of hydrofluoric acid (40%) !!! See Appendix A.	US-Co, UC_2-Fe, (UZr)C-Nb, (UZr)C-Ta, (UZr)C-W (!!!).
Ct m7	Nitric acid (1.40) Sulfuric acid (1.84) Hydrofluoric acid (40%)	10 ml 20 ml 10 ml	Secs to mins. !!! See Appendix A.	NbC_z-NbFe-Nb.
Ct m8	Dist. water Nitric acid (1.40) Hydrochloric acid (1.19)	50 ml 47 ml 3 ml	Secs to mins.	TiC-Ni.
Ct m9	Hydrofluoric acid (40%) Nitric acid (1.40) Lactic acid (90%)	50 ml 50 ml 3 ml	1–5 min. !!! See Appendix A.	UO_2-Nb (!!!).

	Etchant		Conditions	Remarks
Ct m10	Sulfuric acid (1.84) Lactic acid (90%) Glacial acetic acid	10 ml 10 ml 10 ml	Secs to mins.	PuC-Pu (!!!).
Ct m11	a. Dist. water Hydrochloric acid (1.19) Nitric acid (1.40) Hydrofluoric acid (40%) b. Dist. water Nitric acid (1.40)	100 ml 6 ml 2 ml 5 ml 10 ml 10 ml	10 s, 50 °C (120 °F). !!! See Appendix A. 3 min, 25 °C (80 °F). Use (a) first, then (b).	UO_2-Al (!!!).
Ct m12	Lactic acid (90%)		Secs to mins.	$(Y_3Al)C$-Y_3C-Y.
Ct m13	Hydrogen peroxide (30%) Ammonia water	10 ml 10 ml	Secs to mins.	UN-U, UN-W (!!!).

Iron Oxide Layers on Iron

No.	Etchant		Conditions	Remarks
OFe m1	Dist. water Aqueous sol. of nitric acid (1%) Aqueous sol of citric acid (5%) Aqueous sol. of thioglycolic acid (5%)	10 ml 5 ml 5 ml 5 ml	15–60 s. Swab.	Fe_2O_3. Fe_3O_4 and Fe are not attacked.
OFe m2	Aqueous sol. of citric acid (10%) Aqueous sol. of sodium thiocyanate (10%)	5 ml 5 ml	45–90 s. Swab.	Fe_2O_3. Fe_3O_4 is not attacked.
OFe m3	a. Dist. water Formic acid (1.22) b. Dist. water Fluoboric acid	15 ml 5 ml 15 ml 5 ml	5 s. Swab, followed by (b), 2 s.	Fe_3O_4. If Fe_2O_3 is to be etched simultaneously, use OFe m2 first and follow up with OFe m3.
OFe m4	Thioglycolic acid (5%) Aqueous sol. of potassium diphthalate (5%) Aqueous sol. of ammonium citrate (5%) Aqueous sol. of citric acid (5%)	10 ml 5 ml 2 ml 3 ml	30–60 s. Swab.	FeO.

OFe m5	*Electrolytic:* Thioglycolic acid (5%)	10 ml	15 s, 2–4 mA/cm^2, 9 V dc, stainless steel cathode. Add sodium chromate solution only shortly before use.	Fe_3O_4. Fe and Fe_2O_3 are not attacked.
	Aqueous sol. of potassium diphthalate (5%)	5 ml		
	Aqueous sol. of ammonium citrate (5%)	2 ml		
	Aqueous sol of sodium chromate (0.5%)	50 ml		

Appendix A: Suggestions for Handling Hazardous Materials

All chemicals, including many metals and oxides, pose some degree of danger to the human organism. This may come about by ingestion through the respiratory or digestive tracts or by external contact with the skin or eyes. Basically, the same precautions apply to the metallographic laboratory as to all chemical laboratories, except that certain specific areas are particularly critical.

Some significant precautions are:

- *Clearly label* all storage containers.
- Dilute *concentrated chemicals* before disposal and observe all local waste-disposal regulations.
- *Critical substances* (flammable, explosive, toxic, or corrosive) should be stored in approved containers in cool, fireproof, isolated areas.
- *Caustic materials,* such as acids, bases, peroxides, and some salts, should be handled only when wearing protective devices such as safety glasses, rubber gloves, and laboratory coats or aprons. Vapors of such materials are often harmful, too. Actual work should be carried out in an effective fume hood with an additional gas mask if evolution of toxic gases and vapors is suspected.
- When preparing etchants containing *aggressive chemicals* such as sulfuric acid, the chemical should always be added to the solvent (water, alcohol, glycerol, etc.) slowly with gentle stirring. External cooling may also be required if heat evolution is particularly strong.
- *Volatile, flammable, and explosive materials,* such as benzene, acetone, ether, perchlorate, nitrate, etc., should not be heated or kept near open flames.
- When preparing microsections of *toxic materials* such as beryllium, and *radioactive substances* or alloys containing uranium, thorium, and plutonium, a glove box or hot cell must be used.

Particularly hazardous chemicals listed in the etchant compositions (Chapters 2 and 3) and in Appendix B are indicated by (!!!). These deserve additional comments.

- *Perchloric acid* in concentrations exceeding 60% is highly flammable and explosive. This danger is greatly increased by the presence of organic materials or metals such as bismuth, which oxidizes readily. Avoid the higher concentration and heating of these solutions, particularly in electrolytic polishing and etching; never store high-concentration perchloric solutions in plastic containers. When mixing perchloric acid and alcohol, highly explosive alkyl perchlorates may form. Perchloric acid should be added slowly under constant stirring. Keep the temperature of the solution below 35 °C (95 °F) and, if necessary, use a coolant bath. Wearing safety glasses is helpful, but working behind a safety shield is preferable.

- *Mixtures of alcohol and hydrochloric acid* can react in various ways to produce aldehydes, fatty acids, explosive nitrogen compounds, etc. The tendency toward explosion increases with increasing molecule size. Hydrochloric acid content should not exceed 5% in ethanol or 35% in methanol. These mixtures should not be stored.
- *Mixtures of alcohol and phosphoric acid* can result in the formation of esters, some of which are potent nerve poisons. If absorbed through the skin or inhaled, severe personal damage may result.
- *Mixtures of methanol and sulfuric acid* may form dimethylene sulfate, an odorless, tasteless compound that may be fatal if absorbed in sufficient quantities into the skin or respiratory tract. Even gas masks do not offer adequate protection. Sulfates of the higher alcohols, however, are not potentially dangerous poisons.
- *Mixtures of chromium (VI) oxide and organic materials* are explosive. Mix with care and do not store.
- *Lead and lead salts* are highly toxic, and the damage produced is cumulative. Care is also recommended when handling cadmium, thallium, nickel, mercury, and other heavy metals.
- *All cyanide compounds (CN)* are highly dangerous because hydrocyanic acid (HCN) may easily form. They are fast-acting poisons that can cause death, even in relatively low concentrations.
- *Hydrofluoric acid* is a very strong skin and respiratory poison that is hard to control. It should be handled with extreme care, because sores resulting from its attack on skin do not heal readily. Hydrofluoric acid also attacks glass, and fumes from specimens etched in HF solution could easily damage front elements of microscope lenses. Specimens should be rinsed thoroughly and in some cases placed in a vacuum desiccator for one to two hours before examination.
- *Picric acid anhydride* is an explosive.

The references on safety and toxicology in Appendix C contain information on potential poisons, symptoms of poisoning, treatment, and prevention.

Appendix B:
Chemicals Used to Prepare Etchants in Chapters 2 and 3

F = flammable, !!! = toxic, E = explosive, L = liquid, G = gas, C = crystalline, D = density.

Name	Formula	Remarks
Acetic acid	CH_3COOH	!!! (caustic)
Acetylacetone (2,3-pentanedione, diacetylmethane)	$C_5H_8O_2$ ($CH_3COCH_2COCH_3$)	F, L, D 0.972
Aluminum chloride	$AlCl_3$	C
Ammonia	NH_3	!!!, G, D 0.596
Ammonia water	$NH_3 + H_2O$	!!!, L, D 0.91
Ammonium acetate	CH_3COONH_4	C
Ammonium chloride	NH_4Cl	C
Ammonium dicitrate (diammonium hydrogen citrate)	$C_6H_{14}N_2O_7$ $[(NH_4)_2HC_6H_5O_7]$	C
Ammonium ditartrate	$(NH_4)_2C_4H_4O_6$	C
Ammonium hydrogen fluoride	$(NH_4)HF_2$	C
Ammonium paramolybdate (molybdic acid)	$(NH_4)_6Mo_7O_{24} \cdot 4H_2O$	C
Ammonium peroxydisulfate	$(NH_4)_2S_2O_8$	C
Ammonium polysulfide	$(NH_4)_2S_X$	!!!, L
Ammonium thiosulfate	$(NH_4)_2S_2O_3$	C
Argon	Ar	C
Bromine	Br_2	!!! (vapor), L, D 3.11
1-Butanol	$CH_3(CH_2)_3OH$	F, L
Cadmium chloride	$CdCl_2 \cdot H_2O$	!!!, C
Cerium (IV) nitrate	$Ce(NO_3)_4$	C
Chromium (III) oxide	Cr_2O_3	C
Chromium (VI) oxide (chromic acid)	CrO_3	!!! (caustic), C
Citric acid	$C_6H_8O_7 \cdot H_2O$	C
Copper (II) ammonium chloride	$(NH_4)_2[CuCl_4] \cdot 2H_2O$	!!!, C
Copper ammonium persulfate	$[Cu(NH_3)_4]S_2O_8$	C
Copper (II) chloride	$CuCl_2 \cdot H_2O$	!!!, C
Copper (II) nitrate	$Cu(No_3)_2 \cdot 6H_2O$	!!!, C
Copper (II) sulfate	$CuSO_4 \cdot 5H_2O$	!!!, C

1,2-ethanediol (dihydroxy ethane, ethylene glycol, glycol)	$C_2H_6O_2$ ($HOCH_2CH_2OH$)	L, D 1.11
Ethanethiol	$C_6H_{14}O_2$	L, D 0.90
Ethanol	C_2H_5OH	F, L, D 0.81–0.79
Ethylene glycol	(See 1,2-ethanediol)	
Fluoboric acid	HBF_4	!!! (caustic), L, D 1.23
Formic acid	HCOOH	L, D 1.22
Glycerol (glycerine)	$C_3H_8O_3$ ($HOCH_2CHOHCH_2OH$)	L, D 1.26
Gold (III) chloride	$AuCl_3 \cdot H_2O$	C
Hydrochloric acid	HCl	!!! (caustic), L, D 1.19
Hydrofluoric acid	$HF + H_2O$	!!! (caustic), L, 40%
Hydrogen	H_2	E, F, G
Hydrogen peroxide	H_2O_2	L, D 1.11
Hydrogen sulfide	H_2S	!!!, G
Iron (III) chloride	$FeCl_3 \cdot 6H_2O$	C
Iron (III) nitrate	$Fe(NO_3)_3 \cdot 9H_2O$	C
Iron (II) sulfate	$FeSO_4 \cdot 7H_2O$	C
Lactic acid	$C_3H_6O_3$	L, D 1.21
Lead acetate	$Pb(CH_3COO)_2$	!!!, C
Magnesium oxide (magnesia)	MgO	C
Mercury (II) nitrate	$Hg(NO_3)_2 \cdot 8H_2O$	!!!, C
Methanol	CH_3OH	!!! L, D 0.76
Nitric acid	HNO_3	!!! (caustic), L, D 1.19
Nitrogen	N_2	G
Oxalic acid	$C_2H_2O_4 \cdot 2H_2O$	!!!, C
Perchloric acid	$HClO_4$	!!! (caustic), L, E, D 1.67
Phosphoric acid	H_3PO_4	!!! (caustic), L, D 1.71
Picric acid	$C_6H_3N_3O_7$	!!! (caustic), E, C
Potassium bicarbonate	$KHCO_3$	C
Potassium carbonate	K_2CO_3	C
Potassium chloride	KCl	C
Potassium cyanide	KCN	!!!, C
Potassium dichromate	$K_2Cr_2O_7$	!!! (caustic), C
Potassium ferricyanide	$K_3[Fe(CN)_6]$	C
Potassium ferrocyanide	$K_4[Fe(CN)_6]$	C

Potassium hydrate solution	$KOH+H_2O$	!!!(caustic), L
Potassium hydrogen fluoride	KHF_2	C
Potassium hydrogen sulfate	$KHSO_4$	C
Potassium hydroxide	KOH	!!! (caustic), C
Potassium iodide	KI	C
Potassium metabisulfite	$K_2S_2O_5$	C
Potassium nitrate	KNO_3	C
Potassium phthalate (di-)	$C_8H_4K_2O_4$	C
Potassium thiocyanate	KSCN	!!!, C
Silver cyanide	AgCN	!!!, C
Silver nitrate	$AgNO_3$	C
Sodium bicarbonate	$NaCHO_3$	C
Sodium carbonate	$Na_2CO_3 \cdot 1OH_2O$	C
Sodium chloride	NaCl	C
Sodium chromate	Na_2CrO_4	C
Sodium cyanide	NaCN	!!!, C
Sodium dichromate	$Na_2Cr_2O_7 \cdot 2H_2O$	!!! (caustic), C
Sodium fluoride	NaF	C
Sodium hydrogen phosphate	$Na_2HPO_4 \cdot 12H_2O$	C
Sodium hydroxide	NaOH	!!! (caustic), C
Sodium sulfate	$Na_2SO_4 \cdot 10H_2O$	C
Sodium sulfate, anhydrous	Na_2SO_4	C
Sodium sulfide	Na_2S	C
Sodium tetraborate	$Na_2B_4O_7$	C
Sodium thiocyanate	NaSCN	C
Sodium thiosulfate (fixer)	$Na_2S_2O_3 \cdot 5H_2O$	C
Spirits of ammonia	NH_3+H_2O	!!!, L, D 0.91
Sulfuric acid	H_2SO_4	!!! (caustic), L, D 1.84
Tartaric acid	$C_4H_6O_6$	L
Thioglycolic acid	$HSCH_2COOH$	L
Thiourea	$CS(NH_2)_2$	C
1,3-dimethyl 2-thiourea	$C_3H_8N_2S$ ($CH_3NHCSNHCH_3$)	C
Tin (II) chloride	$SnCl_2 \cdot 2H_2O$	!!! (caustic), C
Vogel's special reagent (stainless steel etchant)	Mixture of tar and sulfurous acid, boiled and filtered; protected trade product	L
Wetting agents	Additives for lowering surface tension	
Zinc chloride	$ZnCl_2$	!!! (caustic), C

Appendix C:
References

The references listed in this appendix are suggested for obtaining a better understanding. Only selected texts are represented; not all sources used to prepare the manual are listed.

Safety and Toxicology

1. W. Braun, A. Dönhardt, *Poisoning Register* (in German) *Vergiftungsregister,* Georg Thieme Verl., Stuttgart, 1970.
2. L. V. Cralley, L. J. Cralley, G. D. Clayton, J. A. Jurgiel, Ed., *Industrial Environmental Health,* The Worker and the Community, Academic Press, New York and London, 1972.
3. A. Hamilton, H. L. Hardy, *Industrial Toxicology,* Publishing Sciences Group, Inc., Acton, Mass., 1974.
4. A. Loomis, *Essentials of Toxicology,* Lea & Febinger, Philadelphia, 1974.
5. F. A. Patty, Ed., *Industrial Hygiene and Toxicology,* Interscience Publishers, New York, 1963.
6. N. I. Sax, *Dangerous Properties of Industrial Materials,* Van Nostrand Reinhold Co., New York/Cincinnati/Toronto/London/Melbourne, 1975.
7. G. Sorbe, *Toxins and Explosives* (in German) *Gifte und explosive Substanzen,* Berufskundliche Reihe zur Fachzeitschrift Chemie für Labor und Betrieb, Bd. 7. Umschau Verl., Frankfurt am Main, 1968.
8. P. G. Stecher, *The Merck Index of Chemicals and Drugs,* Merck & Co., Inc., Rahway, N. J. (Get newest edition. Useful for identification of unknown materials).
9. H. E. Stockinger, Ed., *Beryllium: Its Industrial Hygiene Aspects,* Academic Press, New York/London, 1966.
10. C. Xinteras, B. C. Johnson, I. de Groot, Ed., *Behavioral Toxicology,* Early Detection of Occupational Hazards, U. S. Department of Health, Education, and Welfare. Public Health Service—Center for Disease Control—National Institute for Occupational Safety and Health (NIOSH), Rockville, Md., 1974.
11. *Federal Controls on Occupational Exposures to Beryllium:* A Rapid Reference Compliance Guide. Rules and Regulations. Federal Register, Vol. 36, No. 157, Aug 13, 1971.
12. *Guide for Safety in the Chemical Laboratory,* Manufacturing Chemists Assn., Van Nostrand Reinhold Co., New York/Cincinnati/Toronto/London/Melbourne, 1972.
13. *Guidelines for Chemical Laboratories* (in German)
Richtlinien für chemische Laboratorien, Laboratoriumsrichtlinien Ausgabe 1972, Hauptverband der gewerblichen Berufsgenossenschaften, Zentralstelle für Unfallverhütung, Bonn.
14. *Protection Against Dangerous Materials* (in German)
Schutz gegen gefährliche Stoffe, Sammlung der Unfallverhütungsvorschriften, Nr. VBG

1a, Ausg. Dez. 1965/März 1969, Hauptverband der gewerblichen Berufsgenossenschaften, C. Heymanns Verl. KG, Köln.

15. *Registry of Toxic Effects of Chemical Substances,* U. S. Department of Health, Education, and Welfare. Public Health Service — Center for Disease Control — National Institute for Occupational Safety and Health (NIOSH), Rockville, Md., June 1976. (Published every year.)

16. *The Industrial Environment — Its Evaluation & Control,* U. S. Department of Health, Education, and Welfare. Public Health Service — Center for Disease Control — National Institute for Occupational Safety and Health (NIOSH), Rockville, Md., 1973.

17. There is a whole list of information available on individual subjects from NIOSH Standards Library, U. S. Department of Health, Education, and Welfare, Public Health Service — Center for Disease Control — National Institute for Occupational Safety and Health (NIOSH), Rockville, Md.
    a) Technical
    b) Surveys
    c) Research Reports
    d) Full Criteria Documents
    e) Mini Criteria Documents
    f) NIOSH Miscellaneous.

General Textbooks and Reviews

1. D. G. Brandon, *Modern Techniques in Metallography,* Butterworth Publ. Co., London, 1966.

2. H. E. Exner, Ed., "Quantitative Analysis of Microstructures in Medicine, Biology and Materials Development," Special Issue 5 of *Practical Metallography,* Dr. Riederer-Verlag GmbH, Stuttgart, 1975.

3. R. M. Fulrath, J. A. Pask, Ed., *Ceramic Microstructures,* John Wiley and Sons, Inc., New York/London/Sydney, 1968.

4. R. C. Gifkins, *Optical Microscopy of Metals,* American Elsevier Publishing Co., Inc., New York, 1970.

5. H. Gleiter, *Structure and Properties of High-Angle Grain Boundaries in Metals* (in German)
   Struktur und Eigenschaften von Grosswinkelkorngrenzen in Metallen, Materialkundlich-Technische Reihe. G. Petzow, Ed., Gebr. Bornträger Verlag, Berlin/Stuttgart, 1977.

6. E. Hornbogen, G. Petzow, "Metallography" (in German)
   "Metallographie," *Zeitschrift für Metallkunde,* Vol. 61 (1970), pp. 81–94.

7. W. Jähnig, *Metallography of Cast Alloys* (in German)
   Metallographie der Gusslegierungen, VEB Deutscher Verlag für Grundstoffindustrie, Leipzig, 1971.

8. E. Kauczor, *Metal in the Microscope. Introduction to Metallographic Microstructures* (in German)
   Metall unter dem Mikroskop, Einführung in die Metallographische Gefügelehre. Fertigung und Betrieb, Vol. 3. H. Determann, W. Malmberg, Ed., Springer Verlag, Berlin/Heidelberg/New York, 1974.

9. H. Modin, *Metallurgical Microscopy,* Butterworth Publ. Co., London, 1973.

10. V. A. Phillips, *Modern Metallographic Techniques and Their Application,* Wiley Interscience, New York/London/Sydney/Toronto, 1971.

11. E. P. Poluschkin, *Structural Characteristics of Metals,* Elsevier Publ. Co., Amsterdam, 1964.

12. J. H. Richardson, "Optical Microscopy for the Materials Sciences," *Monographs and Textbooks in Materials Science,* Vol. 3, Marcel Dekker, Inc., New York and Basel, 1970.

13. W. Rostocker, J. R. Dvorak, *Interpretation of Metallographic Structures,* Academic Press, New York, London, 1965.

14. H. Schumann, *Metallography* (in German)
Metallographie, VEB Deutscher Verlag für Grundstoffindustrie, Leipzig, 1967.

15. R. E. Smallman, *Modern Physical Metallurgy,* Butterworth International, Woburn, Mass., 1970.

16. C. S. Smith, *A History of Metallography,* University of Chicago Press, 1960.

17. H. Thompson, *Microscopical Techniques in Metallurgy,* Sir Isaac Pitman & Sons, Ltd., London, 1954.

18. *Applications of Modern Metallographic Techniques,* American Society for Testing and Materials (ASTM) Special Technical Publ. No. 480, Symposium Philadelphia, 1966.

19. *Fifty Years of Progress in Metallographic Techniques,* American Society for Testing and Materials (ASTM) Special Technical Publ. No. 430, Symposium Atlantic City 1966, Philadelphia.

20. *Symposium on Metallography in Color,* American Society for Testing and Materials (ASTM) Special Technical Publ. No. 86, Symposium Detroit, 1948, Philadelphia.

Metallographic Preparation

1. M. Beckert, H. Klemm, *Handbook of Metallographic Etching Techniques* (in German)
Handbuch der metallographischen Ätzverfahren, VEB Deutscher Verlag für Grundstoffindustrie (2nd Ed.), Leipzig, 1976.

2. A. F. Bogenschütz, *Etching Semiconductors* (in German)
Ätzpraxis für Halbleiter, Carl Hanser Verlag, München, 1967.

3. P. M. French, J. L. McCall, Ed., *Interpretive Techniques for Microstructural Analysis,* Plenum Press, New York, 1977.

4. H. Freund, Ed., *Metallographic Specimen Preparation for Optical Microscopy — Handbook for Technical Microscopy,* Vol. III/Part I (in German)
Metallographische Probenpräparation für die mikroskopische Untersuchung — Handbuch der Mikroskopie in der Technik, Bd. III/Teil 1, Umschau Verlag, Frankfurt am Main, 1968.

5. R. Galopin, N.F.M., *Microscopic Study of Opaque Minerals,* Henry W. Heffer and Sons Ltd., Cambridge, 1972.

6. G. L. Kehl, *The Principles of Metallographic Laboratory Practice,* Metallurgy and Metallurgical Engineering Series, McGraw-Hill Book Co., New York, 1949.

7. H. E. Knechtel, W. F. Kindle, J. L. McCall, R. D. Buchheit, *Tools and Techniques in Physical Metallurgy,* Vol. 1, F. Weinberg (Ed.); pp. 321–399, M. Dekker Inc., New York, 1970.

8. J. L. McCall, W. M. Mueller, Ed., *Microstructural Analysis, Tools and Techniques,* Plenum Press, New York, 1973.

9. J. L. McCall, W. M. Mueller, Ed., *Metallographic Specimen Preparation,* Plenum Press, New York, 1971.

10. J. A. Nelson, "Metallographic Techniques in the Electronics Industry," *Metallography,* Vol. 9, No. 2, American Elsevier Publishing, 1976, pp. 109–122.

11. J. A. Nelson, E. D. Albrecht, "The Basics of Metallography" (Parts I and II), *Heat Treating*, Apr/June 1976, pp. 19–23.

12. L. E. Samuels, *Metallographic Polishing by Mechanical Methods*, Pitman Publ. Corp., London and Melbourne, 1971.

13. Mc. G. Tegart, *The Electrolytic and Chemical Polishing of Metals in Research and Industry*, Pergamon Press, London, 1959.

14. B. Tuck, "The Chemical Polishing of Semiconductors," *J. Mat. Science*, Vol. 10, 1957, pp. 321–339.

15. R. S. Williams, V. O. Homerberg, *Principles of Metallography*, Metallurgy and Metallurgical Engineering Series, McGraw-Hill Book Co., New York, 1948.

16. J. L. Woodbury, E. D. Albrecht, "The Reflective Decision Maker," *Industrial Research*, May 1976, pp. 78–81.

17. *ANALYST: Preparation of Metallographic, Ceramic, and Petrographic Samples*, Buehler Ltd. and A. I. Buehler Inc., Evanston, Ill.

18. "Metallographic Reagents for Iron and Steel," *Metal Progress Data Book*, Vol. 106, No. 1, American Society for Metals, Metals Park, Ohio, 1974.

19. "Metallographic Sample Preparation," *Metal Digest*, Buehler Ltd., Vol. 11, No. 2/3.

20. *METALOG: Preparation of Metallographic, Ceramic, and Petrographic Samples*, Struers Scientific Instruments, Copenhagen, Denmark.

21. "Petrographic Sample Preparation," *Metal Digest*, Buehler Ltd., Vol. 12/13, No. 1.

22. *1973 Annual Book of ASTM Standards*, Part 31, American Society for Testing and Materials, Philadelphia, 1973.

23. *1975 Annual Book of ASTM Standards*, Part 11: Metallography; Nondestructive Tests, American Society for Testing and Materials, Philadelphia, 1975.

Handbooks, Compilations, and Tables

1. R. J. Gray, J. L. McCall and others (Ed.), *Microstructural Science*, Vol. 1–5, Proceedings of Annual Meetings of the International Metallographic Society, American Elsevier Publ. Co., Inc., New York, 1974–1977.

2. L. Habraken, J. L. deBrouwer, *Basic Metallography* (in German)
De Ferri Metallographia, Grundlagen der Metallographie, Teil I, Presses Académiques Européennes, Bruxelles, 1966.

3. H. Hanemann, A. Schrader, *Atlas of Microstructures*, Vol. I Iron-Carbon Steels (1933), Vol. II Cast Irons (1936), Vol. III Binary Aluminum Alloys (1952) (in German)
Atlas Metallographicus, Gebr. Bornträger Verlag, Braunschweig.

4. V. Horn in cooperation with W. Bernhardt, K. Buness, E. Kretzschmar and H. Stein, *Atlas of Microstructures in Welding* (in German)
Schweisstechnischer Gefügeatlas, VEB Verlag Technik, Berlin, 1974.

5. F. Jeglitsch, G. Petzow, "Progress in Metallography," *Practical Metallography*, Special Edition 4 (in German)
"Fortschritte in der Metallographie," Dr. Riederer-Verlag GmbH, Stuttgart, 1975.

6. R. Mitsche, F. Jeglitsch, G. Petzow, "Progress in Metallography," *Practical Metallography*, Special Edition 3 (in German)
"Fortschritte in der Metallographie," Dr. Riederer-Verlag GmbH, Stuttgart, 1972.

7. F. K. Naumann, *The Book of Failures* (in German)
Das Buch der Schadensfälle, Dr. Riederer-Verlag GmbH, Stuttgart, 1976.

8. J. Orlich, A. Rose, P. Wiest, *Atlas for Heat Treatment of Steels* (in German) *Atlas zur Wärembehandlung der Stähle,* Band 3, Verlag Stahleisen, Düsseldorf, 1973.

9. A. Schrader, A. Rose, *Microstructures of Steels* (in German) *De Ferri Metallographia,* Teil II, Gefüge der Stähle, Verlag Stahleisen, Düsseldorf, 1966.

10. R. W. Wilson, *Metallurgy and Heat Treatment of Tool Steels,* McGraw-Hill Company (U.K.) Ltd., Maidenhead, Berkshire, 1975.

11. "Atlas of Microstructures of Industrial Alloys," *Metals Handbook,* 8th Ed., Vol. 7, American Society for Metals, Metals Park, Ohio, 1972.

12. "Fractography and Atlas of Fractographs," *Metals Handbook,* 8th Ed., Vol. 9, American Society for Metals, Metals Park, Ohio, 1974.

13. "Metallography, Structures and Phase Diagrams," *Metals Handbook,* 8th Ed., Vol. 8, American Society for Metals, Metals Park, Ohio, 1973.

14. *Source Book in Failure Analysis,* American Society for Metals, Metals Park, Ohio, 1974.

15. *Transformation and Hardenability in Steels,* Climax Molybdenum Corp., Ann Arbor, Mich., 1967.

Journals of Mostly Metallographic Content

1. *Metallography.* An International Journal, English, 4 issues/year. Since 1968. American Elsevier Publishing Co., Inc., 52 Vanderbilt Ave., New York, N.Y. 10017. In Cooperation with the International Metallographic Society.

2. *Microstructural Science.* Proceedings of the Annual Technical Meetings of the International Metallographic Society. Since 1972. American Elsevier Publishing Co., Inc., 52 Vanderbilt Ave., New York, N. Y. 10017.

3. *Praktische Metallographie/Practical Metallography.* Bilingual (German-English). 12 issues/year. Detailed comprehensive index for each year, also every ten years. Since 1964. Dr. Riederer-Verlag GmbH, 60 Johannesstrasse, 7000 Stuttgart 1, Germany. In Cooperation with the German Society for Metals (Deutsche Gesellschaft für Metallkunde).

4. *Special Edition Series of Practical Metallography.* Lecture Compilations of Metallographic Meetings. Mostly in German. Published irregularly since 1970. Dr. Riederer-Verlag GmbH, 60 Johannesstrasse, 7000 Stuttgart 1, Germany.

Appendix D:
Some Suppliers of Metallographic Equipment and Materials

NORTH AMERICA

Canada

Anglo Canadian Scientific Co.
P. O. Box 691
Don Mills, Ontario

All materials and equipment for metallographic and ceramographic specimen preparation.

Tech-Met Canada Ltd./Ltee.
31 Progress Ave. 5
Scarborough (Toronto)
Ontario MIP 4S6

All materials and equipment for metallographic, petrographic, and ceramographic specimen preparation. Chemicals, microscopes, and apparatus for microstructural analysis.

Mexico

Tecnicos Argostal, S.A.
Apdo. Postal M-2511
Mexico 1, D.F.

All materials and equipment for metallographic, petrographic, and ceramographic specimen preparation. Chemicals, microscopes, and apparatus for microstructural analysis.

United States

Buehler Ltd.
2120 Greenwood St.
P. O. Box 1459
Evanston, Ill. 60204

All materials and equipment for metallographic, petrographic, and ceramographic specimen preparation. Chemicals, microscopes, and apparatus for microstructural analysis.

Buehler Ltd.
2525 North Loop West
Suite 124
Houston, Tex. 77008

All materials and equipment for metallographic, petrographic, and ceramographic specimen preparation. Chemicals, microscopes, and apparatus for microstructural analysis.

Buehler Ltd.
9010 Reseda Blvd.
Suite 110
Northridge, Calif. 91324

All materials and equipment for metallographic, petrographic, and ceramographic specimen preparation. Chemicals, microscopes, and apparatus for microstructural analysis.

Max Erb Instrument Co.
2112 W. Burbank Blvd.
Burbank, Calif. 91506

All materials and equipment for metallographic, petrographic, and ceramographic specimen preparation. Microscopes and optical instruments.

Excel Metallurgical
P. O. Box 3838
Springfield, Mass. 01101

All materials and equipment for metallographic, petrographic, and ceramographic specimen preparation. Microscopes and optical instruments.

Fisher Scientific
711 Forbes Ave.
Pittsburgh, Pa. 15219
Offices throughout the U.S.A.

All materials and equipment for metallographic, petrographic, and ceramographic specimen preparation.

T. C. Jarrett Co.
P. O. Box 15397
Denver, Colo. 80215

Complete system in preparing metallurgical and geological specimens.

Leco Corp.
3000 Lakeview Ave.
St. Joseph, Mich. 49085

All materials and equipment for metallographic and ceramographic specimen preparation.

Struers Inc.
20102 Progress Drive
Cleveland, Ohio 44136

All materials and equipment for metallographic and ceramographic specimen preparation. Chemicals, microscopes.

VWR Scientific Inc.
3745 Bayshore Blvd.
Brisbane, Calif. 94005
Offices throughout the U.S.A.

All materials and equipment for metallographic, petrographic, and ceramographic specimen preparation.

Wilkens Anderson
4525 W. Division St.
Chicago, Ill. 60651

All materials and equipment for metallographic, petrographic, and ceramographic specimen preparation.

SOUTH AMERICA

Argentina

Lutz, Ferrando y Cia. S.A.
Florida (R5)
Buenos Aires

All materials and equipment for metallographic and ceramographic specimen preparation.

SIREX
Libertad 836
Buenos Aires

All materials and equipment for metallographic, petrographic, and ceramographic specimen preparation.

Bolivia

GAMMA
Casilla No. 4316
La Paz

All materials and equipment for metallographic, petrographic, and ceramographic sepcimen preparation.

Brazil

Panambra S.A. Av. Senador Queiroz 150 Sao Paulo	All materials and equipment for metallographic and ceramographic specimen preparation.
POLITEC Importacao e Comercia Ltda. Rua Correa de Lemos 309 04140 Sao Paulo	All materials and equipment for metallographic, petrographic, and ceramographic specimen preparation.

Chile

Forestier-Intrumentos Scientificos G. Busch & Cia. Ltda. Casilla 191-V Valparaiso	All materials and equipment for metallographic, petrographic, and ceramographic specimen preparation. Microscopes and optical instruments.
Erwin Schneuer K Casilla 9339 Moneda 1137/85–87 Santiago de Chile	All materials and equipment for metallographic and ceramographic specimen preparation.

Colombia

Instrumentación H. A. Langebaek & Kier S.A. Carrera 7 No. 48–59 Bogota	All materials and equipment for metallographic and ceramographic specimen preparation.
Pantecnica S.A. Apartado Aereo 7189 Bogota	All materials and equipment for metallographic, petrographic, and ceramographic specimen preparation.
Milciades Sanchez Apartado Aereo 4675 Bogota	All materials and equipment for metallographic, petrographic, and ceramographic specimen preparation.

Ecuador

Panandina Cia. Ltda. P. O. Box 3882 Quito	All materials and equipment for metallographic, petrographic, and ceramographic specimen preparation.
SUMITEC Suministros Tecnicos Ltda. P. O. Box 259-4492 Guyaquil	All materials and equipment for metallographic and ceramographic specimen preparation.

Peru

H. W. Kessel Apartado 552 Lima	All materials and equipment for metallographic, petrographic, and ceramographic specimen preparation.

Uruguay

Casa Stapff
Casilla Correo 640
Montevideo

All materials and equipment for metallographic, petrographic, and ceramographic specimen preparation.

Venezuela

Ferrum C.A.
P. O. Box 70.624
Caracas

All materials and equipment for metallographic, petrographic, and ceramographic specimen preparation.

C. Hellmund & Cia. S.A.
Avenida Pantin
Edificio Casa Hellmund
Caracas

All materials and equipment for metallographic and ceramographic specimen preparation.

CENTRAL AMERICA

Dominican Republic

Quimico Industrial
Apartado de Correos No. 2058
Santo Domingo

All materials and equipment for metallographic, petrographic, and ceramographic specimen preparation.

Nicaragua

Roberto Teran G.
Apartado Postal 689
Managua, D.N.

All materials and equipment for metallographic, petrographic, and ceramographic specimen preparation.

Panama

Promed, S.A.
Apartado 6281
Panama 5

All materials and equipment for metallographic, petrographic, and ceramographic specimen preparation.

EUROPE

Austria

Buehler-Met Handels GmbH
Rainerring 22
A-2500 Baden bei Wien

All materials and equipment for metallographic, petrographic, and ceramographic specimen preparation. Chemicals, microscopes, and apparatus for microstructural analysis.

C. Reichert Optische Werke AG
Hernalser Hauptstr. 219
A-1170 Wien

Special metallographic equipment and microscopes.

Belgium

S. A. Analis
Rue Dewez, 14
B-5000 Namur

All materials and equipment for metallographic, petrographic, and ceramographic specimen preparation.

BODSON (Établissements)
6, Quai St. Leonard
B-4000 Liège

All materials and equipment for metallographic and ceramographic specimen preparation.

Denmark

Struers K/S
38 Skindergade
DK-1159 Kopenhagen K

All materials and equipment for metallographic and ceramographic specimen preparation. Chemicals and microscopes.

Finland

AB Axel von Knorringin
Tekn. BYRA
P. O. Box 20
Helsinki 38

All materials and equipment for metallographic, petrographic, and ceramographic specimen preparation.

Kaukomarkkinat Oy
Fabianinkavt 9
Helsinki 13

All materials and equipment for metallographic and ceramographic specimen preparation.

France

ISI
Les Instruments Scientifiques
et Industriels
77, Avenue Parmentier
B. P. 254
75526 Paris

All materials and equipment for metallographic and ceramographic specimen preparation.

Presi
2 Avenue Hector-Berlioz
F Poisat/Grenoble
38310 Eybens

All materials and equipment for metallographic and ceramographic specimen preparation.

Testwell, S.A.
B. P. 176-09
Paris 75

All materials and equipment for metallographic, petrographic, and ceramographic specimen preparation.

East Germany

Jenoptik Jena GmbH
Carl-Zeiss-Str. 1
DDR 69 Jena

Electropolishers and microscopes.

VEW Rathenower Optische Werke
DDR Rathenow

Grinding and polishing equipment.

West Germany

Buehler-Met GmbH 35 Zeppelinstr. D-7203 Ostfildern (Kemnat)/ Stuttgart	All materials and equipment for metallographic, petrographic, and ceramographic specimen preparation. Chemicals, microscopes, and apparatus for microstructural analysis.
Buehler-Met GmbH Lessingstr. 66/68 D-46 Dortmund	All materials and equipment for metallographic, petrographic, and ceramographic specimen preparation. Chemicals, microscopes, and apparatus for microstructural analysis.
P. F. Dujardin & Co. 16 Wiesenstr. D-4000 Düsseldorf-Heerdt	Materials and equipment for metallographic specimen preparation. Chemicals. Vogel's special reagent.
Elektroschmelzwerk Kempten GmbH P. O. Box 609 D-8000 München 33	Diamond pastes of polycrystalline diamonds.
K. & B. Grubbs Instrument GmbH & Co. KG 15 Froschkönig-Weg D-4000 Düsseldorf	All materials and equipment for metallographic and ceramographic specimen preparation. Chemicals.
Iminex 14 Hauptstr. D-6145 Lindenfels 2	Materials and equipment for metallographic specimen preparation.
Jenoptik Jena GmbH 6 Weender Landstr. D-3400 Göttingen	Electropolishers. Microscopes.
R. Jung P. O. Box 1120 D-6901 Nussloch/Heidelberg	Microtome and micromilling devices.
Kontron GmbH 1 Oskar v. Millerstr. D-8057 Eching/München	Special materials for metallographic specimen preparation. Microscopes and apparatus for microstructural analysis.
Kulzer & Co. GmbH P. O. Box 1749 D-6380 Bad Homburg	Plastics for cold mounting. Metal grinding and polishing system.
Leco Instrumente GmbH Benzstr. 8011 Kirchheim bei München	All materials and equipment for metallographic and ceramographic specimen preparation.
Ernst Leitz GmbH P. O. Box 2020 D-6330 Wetzlar	Etching chamber for evaporating of interference layers. Microscopes.
Multi Metal P. O. Box 2705 D-4150 Krefeld	Plastics for embedding specimens.

Scan Dia P. O. Box 3031 D-5800 Hagen	Materials and equipment for metallographic specimen preparation.
Ernst Winter & Sohn 58 Osterstr. D-2000 Hamburg 21	Polishing equipment, diamond pastes, and tools.
Jean Wirtz 73 Charlottenstr. D-4000 Düsseldorf 1	All materials and equipment for metallographic and ceramographic specimen preparation. Chemicals. Apparatus for microstructural analysis and microscopes.

Great Britain

Banner Scientific Ltd. 3, Three Spires Ave. Coventry CV6 1LE	All materials and equipment for metallographic, petrographic, and ceramographic specimen preparation. Chemicals, microscopes, and apparatus for microstructural analysis.
Metallurgical Services Reliant Works Brockham, Betchworth Surrey/England	Special metallographic materials and equipment.
Vickers Ltd. Vickers Instruments Haxby Rd. York YO3 7SD	All materials and equipment for metallographic and ceramographic specimen preparation.

Greece

Carco-Technica Ltd. 23–24 Eleftherias Sq. Athens 113	All materials and equipment for metallographic, petrographic, and ceramographic specimen preparation.
M. J. Priniotakis 5 Evrydikis Street Athens 516	All materials and equipment for metallographic and ceramographic specimen preparation.

Ireland

Thomas H. Mason & Sons Ltd. Crane Lane (off Dame St.) Dublin 2	All materials and equipment for metallographic and ceramographic specimen preparation.

Italy

Gagliani & sas Via Edolo 19 20125 Milano	All materials and equipment for metallographic, petrographic, and ceramographic specimen preparation.
W. Pabisch S.p.A. Via Borromei 1 B/4 I-20123 Milano	All materials and equipment for metallographic and ceramographic specimen preparation.

Liechtenstein

Balzers AG. High-Vacuum Techn. and Thin Film Techn. FL-9496 Balzers	Apparatus for evaporation of interference layers (Pepperhoff etch).

Netherlands

Lameris Instrumenten N.V. Biltstraat 149 Utrecht	All materials and equipment for metallographic and ceramographic specimen preparation.
VIBA, N.V. Koningin Emmakade 199 The Hague	All materials and equipment for metallographic, petrographic, and ceramographic specimen preparation. Microscopes and optical instruments.

Norway

A/S Christian Falchenberg Nedre Slottsgate 23 Oslo	All materials and equipment for metallographic, petrographic, and ceramographic specimen preparation.
Nerliens Kemisk-Tekniske Aktieselskap Tollbugt. 32 Oslo 1	All materials and equipment for metallographic and ceramographic specimen preparation.

Portugal

Equipamentos de Laboratorio Lda. Apartado 1,100 Lisbon 1	All materials and equipment for metallographic, petrographic, and ceramographic specimen preparation.
Viara Comercail (Máquinas) Lda. Rua Delfim Ferreira, 509 Porto	All materials and equipment for metallographic and ceramographic specimen preparation.

Spain

C. R. Marés, S.A. Calle Valencia 333 Barcelona-9	All materials and equipment for metallographic and ceramographic specimen preparation.
P.A.C.I.S.A. P. O. Box 7023 Madrid 5	All materials and equipment for metallographic, petrographic, and ceramographic specimen preparation.

Sweden

Bergman & Beving AB Karlavägen 76 Stockholm 10	All materials and equipment for metallographic and ceramographic specimen preparation.

Brandt Optik AB P. O. Box 27053 S 102 51 Stockholm	All materials and equipment for metallographic, petrographic, and ceramographic specimen preparation. Microscopes and optical instruments.

Switzerland

Carl Bittmann Herzogenmühlestr. 14 CH-8015 Zürich	All materials and equipment for metallographic and ceramographic specimen preparation.
Buehler-Met AG Zollfreilager Dreispitz CH-4023 Basel	All materials and equipment for metallographic, petrographic, and ceramographic specimen preparation. Chemicals, microscopes, and apparatus for microstructural analysis.

Yugoslavia

HERMES Mosa Pijade 27 61000 Ljubljana	All materials and equipment for metallographic, petrographic, and ceramographic specimen preparation.
Jugolaboratorija Ul. 7 Jula No. 44 11000 Beograd	All materials and equipment for metallographic and ceramographic specimen preparation.

ASIA

Bangladesh

Beas Enterprises GPO Box 787 Dacca 2	All materials and equipment for metallographic, petrographic, and ceramographic specimen preparation.

Hong Kong

Schmidt & Co. P. O. Box 297 Hong Kong	All materials and equipment for metallographic, petrographic, and ceramographic specimen preparation.

India

J. T. Jagtiani National House Tulloch Rd. Apollo Bunder Bombay 1	All materials and equipment for metallographic and ceramographic specimen preparation.

Indonesia

P. T. Ahrend Indonesia Djalan Kapten P. Tendean 19 (Hegarmanah Kulon) Bandung	All materials and equipment for metallographic and ceramographic specimen preparation.
Schmidt Scientific (Pte) Ltd. Katong P. O. Box 76 Singapore 15	All materials and equipment for metallographic, petrographic, and ceramographic specimen preparation.

Japan

Kasai Shoko Co. 5-6-37 Tsanashimahigashi Kohoku-ku Yokohama 223	All materials and equipment for metallographic and ceramographic specimen preparation.
Marumoto Kogyo Co. 2-1 Kyobashi Chuo-ku Tokyo 104	All materials and equipment for metallographic and ceramographic specimen preparation. Standard specimens for metallography (for example, grain size, carbon content).
Maruto Co. 1-1-10 Yushima Bunkyo-ku Tokyo 113	Precise cutting machines. Apparatus for determination of the crystal orientation by etch figures.
Rigaku Denki Co. 2-9-8 Sotokanda Chiyoda-ku Tokyo 101	Portable metallography apparatus.
Sankei Co. Ltd. Tsunashima Dai-2 Bldg. No. 20-12, 3-Chome, Jushima Bunkyo-ku Tokyo	All materials and equipment for metallographic, petrographic, and ceramographic specimen preparation. Chemicals, microscopes, and apparatus for microstructural analysis.
Yamamoto Kagaku Kogu Kenkyusha 2-15-4 Sakaemachi Hunabashi Chiba 273	Standard specimens for metallography (for example, grain size, carbon content, hardness).
Yunion Kogaku Co. 2-20-9 Shimura Itabashi-ku Tokyo 174	Optical microscopes. Abrasives and polishing machines.

Malaysia

Jebsen & Jessen (M) Sedn. Bhd. Denmark House 84 Jalan Ampong Kuala Lumpur	All materials and equipment for metallographic and ceramographic specimen preparation.
Schmidt Scientific Sdn. Bhd. P. O. Box 592 Kuala Lumpur 08-03	All materials and equipment for metallographic, petrographic, and ceramographic specimen preparation.

Nepal

Gita Trading Co. GPO Box 479 Katmandu	All materials and equipment for metallographic, petrographic, and ceramographic specimen preparation.

Pakistan

Nazer & Co. Habib Bank Building Victoria/Bunder Road Karachi 3	All materials and equipment for metallographic and ceramographic specimen preparation.

Philippines

Engineering Equip. Inc. P. O. Box 7160 Airmail Exchange Office Manila Int'l. Airport 3120	All materials and equipment for metallographic, petrographic, and ceramographic specimen preparation.
Inhelder-Don Baxter Lab. Inc. 41, Pioneer St. Mandeluyong Rizal D-713	All materials and equipment for metallographic and ceramographic specimen preparation.

Singapore

Jebsen & Jessen (S) Pte. Ltd. 10th Floor Supreme House Penang Road Singapore	All materials and equipment for metallographic and ceramographic specimen preparation.

South Korea

Shin Han Scientific Co. Ltd. Int'l P. O. Box 1250 Seoul	All materials and equipment for metallographic, petrographic, and ceramographic specimen preparation.

Sri Lanka

Laboratory Equipment Co. 3rd Floor, Y.M.B.A. Bldg. Colombo 1	All materials and equipment for metallographic, petrographic, and ceramographic specimen preparation.

Taiwan

San Kwang Instr. Co. Ltd.
No. 20, Yung Sui Rd.
Taipei 100

All materials and equipment for metallographic, petrographic, and ceramographic specimen preparation.

Thailand

Sis Company Ltd.
1248 Nakornchaisri Rd.
Bangkok 3

All materials and equipment for metallographic, petrographic, and ceramographic specimen preparation.

AUSTRALIA

H. B. Selby & Co. Pty. Ltd.
352-368 Ferntree Gully Rd.
Notting Hill, Victoria 3168

All materials and equipment for metallographic and ceramographic specimen preparation.

Watson Victor Ltd.
P. O. Box 100
North Ryde, NSW 2113

All materials and equipment for metallographic, petrographic, and ceramographic specimen preparation.

NEW ZEALAND

Watson Victor Ltd.
Physics & Engin.
P. O. Box 1180
Wellington

All materials and equipment for metallographic, petrographic, and ceramographic specimen preparation.

Geo. W. Wilton & Co. Ltd.
77 Carlton Gore Rd.
New Market
Auckland, C.1

All materials and equipment for metallographic and ceramographic specimen preparation.

MIDDLE EAST

Cyprus

Advanced Technical Services GmbH.
P. O. Box 2349
Nicosia

All materials and equipment for metallographic, petrographic, and ceramographic specimen preparation.

Egypt

Johs. Rieckermann
140, Tehrir St., 6th Floor
Cairo, A.R.E.

All materials and equipment for metallographic, petrographic, and ceramographic specimen preparation.

Iran

Soofer Co. Khark & Shahreza Avenue Teheran	All materials and equipment for metallographic, petrographic, and ceramographic specimen preparation.
IMACO Ltd. Chiaban Sezavar 111 Teheran	All materials and equipment for metallographic and ceramographic specimen preparation.

Iraq

Leon Kouyoumdijan & Co. Fixit House Sa'Adoon St. Alwiya Baghdad	All materials and equipment for metallographic and ceramographic specimen preparation.

Israel

Agentex Ltd. 3 Bograshov Tel-Aviv	All materials and equipment for metallographic and ceramographic specimen preparation.
L. Kardos P. O. Box 11033 Tel-Aviv	All materials and equipment for metallographic, petrographic, and ceramographic specimen preparation.

Jordan

Mouasher Cousins P. O. Box 1387 Amman	All materials and equipment for metallographic, petrographic, and ceramographic specimen preparation.

Lebanon

Comptoir D'Électricité Av. Bechara Khoury Beirut	All materials and equipment for metallographic, petrographic, and ceramographic specimen preparation.

Saudi Arabia

Abdulmajeed Y. Al-Reshaid P. O. Box 363 Al-Khobar	All materials and equipment for metallographic, petrographic, and ceramographic specimen preparation.

Syria

The Arab Trad. & Eng. Office Port Said St. 137 Damascus	All materials and equipment for metallographic, petrographic, and ceramographic specimen preparation.

Technical & Laboratory
Appliances
Fardoss Street 79
Kassas & Sadate Building
Damascus

All materials and equipment for metallographic and ceramographic specimen preparation.

Turkey

Alemdar Enternasyonal A.S.
Veli Alemdar Han 717
Karakoy, Istanbul

All materials and equipment for metallographic and ceramographic specimen preparation.

AFRICA

Algeria

Scientifix
3, Boulevard Colonel
Amirouche
Alger

All materials and equipment for metallographic and ceramographic specimen preparation.

Angola

Equipamentos Tecnicos Lda.
Rua Serpa Pinto 391
Caixa Postal 6319
Luanda

All materials and equipment for metallographic, petrographic, and ceramographic specimen preparation.

Morocco

Chimilabo
193, Avenue des Forces
Armées Royales
Casablanca

All materials and equipment for metallographic, petrographic, and ceramographic specimen preparation.

Jacques Guy-Moyat
Passage Sumica
Casablanca

All materials and equipment for metallographic and ceramographic specimen preparation.

Mozambique

Artur Ballossini
P. O. Box 288
Lourenco Marques

All materials and equipment for metallographic, petrographic, and ceramographic specimen preparation.

Telecom-Equipal
Telecomunicacoes de Mocambique
C. Postal 310
Maputo

All materials and equipment for metallographic, petrographic, and ceramographic specimen preparation.

Rhodesia

Taeuber & Corssen (Pty) Ltd.
P. O. Box 3190
Salisbury

All materials and equipment for metallographic, petrographic, and ceramographic specimen preparation.

South Africa

OPTOLABOR
3rd Floor, Standard House
40 de Korte St.
Braamfontain
Johannesburg

All materials and equipment for metallographic and ceramographic specimen preparation.

T & C Scientific
Suppliers (Pty) Ltd.
P. O. Box 1366
Johannesburg

All materials and equipment for metallographic, petrographic, and ceramographic specimen preparation.

Zambia

Baird & Tatlock
(Zambia) Ltd.
P. O. Box 1097
Brunell Rd.
Ndola

All materials and equipment for metallographic, petrographic, and ceramographic specimen preparation.

Index

$A_{III}B_V$ and $A_{II}B_{VI}$ compounds, 68-69
Aluminum and aluminum alloys, 39-43
Amalgamates, 71
Antimony and antimony alloys, 48-49

Bearing alloys, 81, 84, 85
Bearing metals, 81
Beryllium and beryllium alloys, 46-48
Bibliographic references, 108-112
 general textbooks and reviews, 109-110
 handbooks, compilations and tables, 111-112
 journals of mostly metallographic content, 112
 metallographic preparation, 110-111
 safety and toxicology, 108-109
Bismuth and bismuth alloys, 48-49
Borides, 98
Brasses, 58-61
Bronzes
 aluminum bronzes, 58-61
 beryllium bronzes, 59
 copper bronzes, 58-61
 lead bronzes, 80
 silicon bronzes, 58
 special bronzes, 59
 tin bronzes, 59, 60

Cadmium and cadmium alloys, 50-51
Carbides, 94-96
Cast iron, 61-68
Cathodic vacuum etching, 29
Cemented carbides, 52, 53, 78
Cerium, 83
Cermets, 99-100
Chemical polishing solutions, 20
Chemicals used to prepare etchants in Chapters 2 and 3, 105-107
Chromium and chromium alloys, 54-57
Cleaning, 23-24
 drying after, 24
 rinsing, 28
 ultrasonic cleaning, 24
Cobalt and cobalt alloys, 51-53
Copper and copper alloys, 58-61
Corrosion-resistant steels, 61-68
Current-voltage characteristics, 18-19
Cutting tools, cobalt-base, 52, 53

Drying of specimens after cleaning, 24
Duralumin, 41
Dysprosium, 83

Electrolytes, 16-18
Electrolytic polishing, 16-20
Erbium, 83
Etching, 24-34
 anodic etching (anodizing), 25, 28, 31, 32 (see also *electrolytic etching*)
 attack-polishing, 21, 31, 32
 cathodic etching, 31, 32
 chemical etching (see *electrochemical etching*)
 cold etching, 31, 32
 controlled etching, 28, 31, 32
 conventional (classical) chemical etching, 25, 27, 30
 crystal-figure etching, 31, 32
 deep etching, 31, 32
 dislocation etching, 31, 32
 dissolution etching, 32, 33
 double etching, 32,33
 drop etching, 32, 33
 drop etching, 32, 33
 dry etching, 32, 33
 electrochemical etching, 25-29, 33
 electrolytic etching, 28, 30 (see also *anodic etching*)
 etch rinsing, 32, 33
 eutectic-cell etching, 32, 33
 grain-boundary etching, 32, 33
 grain-contrast etching, 32, 33
 heat tinting, 27, 32, 33
 hot etching, 32, 33
 identification etching, 32, 33
 immersion etching, 32, 33
 immersion etching, cyclic, 32, 33

Etching *(continued)*
ion etching, 25, 29, 30
long-term etching, 32, 33
macroetching, 32, 33
microetching, 32, 33
multiple etching, 32, 33
network etching, 32, 33
optical etching, 25, 33
physical etching, 25, 29, 33
potentiostatic etching, 25, 28, 30, 33
precipitation etching, 27, 32, 33
primary etching, 32, 34
printing, 32, 34
reproducible etching, 30
secondary etching, 32, 34
segregation etching, 34
short-term etching, 32, 34
shrink etching, 32, 34
staining, 32, 34
strain etching, 34
swabbing, 32, 34
thermal etching, 25, 29, 32, 34
wet etching, 32, 34
wipe etching, 39
Etching nomenclature; explanation of etching terms, 31-34
Etching, post-treatment, 28-29
Europium, 83
Evaporation of interference layers, 25, 29, 30

Flow lines in steel forgings, 66

Gadolinium, 83
German silver, 60
Germanium and germanium alloys, 68-69
Gold and gold alloys, 43-46
Grinding, 8-16, 22
automatic grinding, 22
coarse grinding, 8, 9, 11
fine grinding, 8, 11
mechanical grinding, 9
pre-grinding, 11
pressure, time, velocity, 15
specimen motion in, 15
substrates, 15-16
Grinding compounds, 11-15
Grinding substrate materials, 15

Hafnium and hafnium alloys, 70-71
Hardened steels, 61, 64-67
Hazardous materials, precautions in handling, 103-104
Heat-resistant steels, 61-68
Holmium, 83

Indium and indium alloys, 50-51
Iridium and iridium alloys, 43-46
Iron and iron alloys, 61-68
Iron oxide layers, 100

Lanthanum and lanthanum alloys, 83
Lead and lead alloys, 79
Lutetium, 83

Magnesium and magnesium alloys, 72-74
Magnetic alloys, 52
Manganese and manganese alloys, 75
Marking of specimens for identification, 7-8
Mercury amalgamates, 71, 72
Metal compounds
borides, 98
carbides, 94-96
nitrides, 97
oxides, 92-94
phosphides and sulfides, 98-99
Metallographic equipment and materials, suppliers, 113-127
Africa, 126-127
Asia, 121-124
Australia, 124
Central America, 116
Europe, 116-121
Middle East, 124-126
New Zealand, 124
North America, 113-114
South America, 114-116
Microtome cutting, 8, 16, 23
Molybdenum and molybdenum alloys, 54-57
Monel metal, 61, 76, 77
Mounting of specimens, 3-7
clamping, 3, 4
plastic embedding, 4-6
properties of mounting materials, 6-7
Multiple polishing, 21
Muntz metal, 61

Neodymium, 83
Nickel and nickel alloys, 75-79
Niobium and niobium alloys, 54-57

Nitride metal compounds, 97
Nondestructive metallographic testing, 34-35

Osmium and osmium alloys, 43-46
Oxide metal compounds, 92-94

Palladium and palladium alloys, 43-46
Phosphide and sulfide compounds, 98-99
Phosphorus distribution in steel, 63
Platinum and platinum alloys, 43-46
Plutonium and plutonium alloys, 81-83
Polishing, 8-23
  anodic polishing, 16-21 (see also *electrolytic polishing*)
  attack-polishing, 21, 31, 32
  automatic polishing, 22
  chemical polishing, 20-21
  coarse polishing, 8, 11
  electrolytic lapping, 21-22
  electrolytic polishing, 16-21 (see also *anodic polishing*)
  etch-polishing, 21
  final polishing, 11
  fine polishing, 11
  mechanical polishing, 9-11
  pre-polishing, 11
  pressure, time, velocity, 15
  specimen motion in, 15
  substrates, 15
Polishing compounds, 10-14
Polishing fluids, 14-15
Polishing methods, combinations, 21-22
Polishing methods, evaluation, 22-23
Polishing substrate materials, 15-16
Praseodymium, 83
Precautionary measures
  in handling hazardous materials, 103-104
  with perchloric acid, 17-18
Precious metals, 43-46
Preparation of metals and alloys, 37-90 (see also entries under specific metals)
Preparation of special ceramics and cermets (ceramography), 91-101 (see also entries under *Metal compounds*)
Primary grain structure in steels, 63
Promethium, 83
Pure iron, 64, 66, 67

Rare earth metals, 83
Redox processes, 25-26
Rhenium and rhenium alloys, 54-57
Rhodium and rhodium alloys, 43-46
Ruthenium and ruthenium alloys, 43-46

Samarium, 83
Selenium and selenium alloys, 68-69
Semiconductors, 68-69
Silicon and silicon alloys, 68-69
Silver and silver alloys, 37-39
Silver solders, 38, 39
Solder alloys containing cadmium, 50
Special steels, 63
Specimen preparation, general, 1
Specimen sectioning, 2-3
Specimen storage, 29-30
Stainless steels, 64-67
Steels, 61-68
Stellite, 51, 53
Structural steels, 61-68
Sulfide inclusions in steel, 63
Sulfide-metal compounds, 98-99
Superalloys, 52, 53, 76-79

Tantalum and tantalum alloys, 54-57
Taper sectioning of specimens, 5-6
Tellurium and tellurium alloys, 68-69
Terbium, 83
Thallium and thallium alloys, 50, 51
Thorium and thorium alloys, 81-83
Thulium, 83
Tin and tin alloys, 84-85
Titanium and titanium alloys, 85-88
Tombac, 61
Tool steels, 64-66
Tungsten and tungsten alloys, 54-57
Type metal, 80, 81

Ultrasonic cleaning, 24
Uranium and uranium alloys, 81-83

Vacuum impregnation, 5
Vanadium and vanadium alloys, 54-57

White metals (babbitts), 80, 81, 84, 85

Ytterbium, 83

Zinc and zinc alloys, 88-90
Zinc platings, 90
Zircaloy, 70, 71
Zirconium and zirconium alloys, 70-71